Handbook of Animal Husbandry

Handbook of Animal Husbandry

Rajesh Kumar

RANDOM PUBLICATIONS
NEW DELHI (INDIA)

Handbook of Animal Husbandry

ISBN 978-93-5111-486-4

Published in 2015 in India by

RANDOM PUBLICATIONS

4376-A/4B, Gali Murari Lal, Ansari Road
New Delhi-110 002
Phone : +9111-43580356, 011-23289044, 011-43142548
e-mail: sales@randompublications.com,
info@randompublications.com, randomexports@gmail.com

Reprinted 2025

Type Setting by : Friends Media, Delhi-110089
Digitally Printed at : Replika Press Pvt. Ltd.

Preface

Animal husbandry is the art and science of taking care of domestic animals that are used primarily as food or product sources. In many places throughout the world, people are essentially specialists in animal husbandry by means of being farmers, ranchers, sheepherders, or anyone who takes care of a variety of animals. A large number of farmers depend on animal husbandry for their livelihood. In addition to supplying milk, meat, eggs, and hides, animals, mainly bullocks, are the major source of power for both farmers and drayers. Thus, animal husbandry plays an important role in the rural economy.

During the past three decades, a rapid global expansion in production and consumption of animal products has led to a so-called "livestock revolution", driven by population and income growth coupled with urbanization. Cheap, often subsidized feed grain, cheap fuel and rapid technological change, particularly in poultry, pork and dairy production, have accelerated the sector's growth to such an extent that it is expected to provide 50 percent of global agricultural output in value terms in the next ten years. Rapid structural change has been associated closely with this growth process in developed and rapidly growing developing countries. Large-scale commercial production, based mostly on feed grain and often globally connected, has emerged to provide growing urban markets with produce.

Handbook of Animal Husbandry provides a comprehensive introduction to the production and rearing of livestock. Topics such as modes of livestock farming, housing methods, feeding techniques, health management, applications of technology, etc, are described in detail. The contents of the book will be immensely useful to students, researchers, farmers, policy makers and all who are related with animal husbandry.

Author

Contents

1

Animal Husbandry: Status and Trends

Livestock contribute 40 percent of the global value of agricultural output and support the livelihoods and food security of almost a billion people. The livestock sector is one of the fastest growing parts of the agricultural economy, driven by income growth and supported by technological and structural change. The growth and transformation of the sector offer opportunities for agricultural development, poverty reduction and food security gains, but the rapid pace of change risks marginalizing smallholders, and systemic risks to the environment and human health must be addressed to ensure sustainability.

In many developing countries, livestock keeping is a multifunctional activity. Beyond their direct role in generating food and income, livestock are a valuable asset, serving as a store of wealth, collateral for credit and an essential safety net during times of crisis. Livestock are also central to mixed farming systems. They consume waste products from crop and food production, help control insects and weeds, produce manure for fertilizing and conditioning fields and provide draught power for ploughing and transport. In some areas, livestock perform a public sanitation function by consuming waste products that would otherwise pose a serious pollution and public health problem.

At the global level, livestock contribute 15 percent of total food energy and 25 percent of dietary protein. Products from livestock provide essential micronutrients that are not easily obtained from plant-based foods.

Almost 80 percent of the world's undernourished people live in rural areas and most depend on agriculture, including livestock, for their livelihoods. Data from the FAO database on Rural Income Generating Activities (RIGA) show that, in a sample of 14 countries, 60 percent of rural households keep livestock. A significant share of the livestock outputs of rural households is sold, making a sizeable contribution to household cash income. In some countries, the poorest rural households are more likely to hold livestock than wealthier ones; although the average number of livestock per household is quite small, this makes livestock an important entry point for poverty alleviation efforts.

Women and men typically face different livelihood opportunities and constraints in managing livestock. Small livestock keepers, particularly women, face many challenges, including: poor access to markets, goods, services and technical information; periodic drought and disease; competing resource uses; policies that favour larger-scale producers or external markets; and weak institutions. Knowledge about, and responsibilities for, various aspects of animal husbandry and livestock production commonly differ between women and men and between age groups. For example, a woman might be responsible for preventing or treating illness in the household's livestock, a man for milking or marketing, boys for grazing or watering, and girls for providing fodder to stall-fed animals.

Poor people, especially young children and their mothers in developing countries, are not consuming enough animal-based food, while other people, particularly in developed countries, are consuming too much. However, high rates of undernourishment and micronutrient deficiency among the rural poor suggest that, despite often keeping livestock, the rural poor consume very little animal-based food. About 4–5 billion people in the world are deficient in iron, which is essential especially for the health of pregnant and lactating women and for the physical and cognitive development of young children. This and other important nutrients are more readily available in meat, milk and eggs than in plant-based foods. Increasing access to affordable animal-based foods could thus significantly improve nutritional status and health for many poor people. However, excessive consumption of livestock products is associated with increased risk of obesity, heart disease and other non-communicable diseases. Furthermore, the rapid growth of the livestock sector means that competition for land and other productive resources puts upward pressure on prices for staple grains as well as negative pressures on the natural-resource base, potentially reducing food security.

Powerful forces of economic change are transforming the livestock sector in many rapidly growing developing countries. Production of livestock, especially pigs and poultry, is becoming more intensive, geographically concentrated, vertically integrated and linked with global supply chains. Higher animal-health and food-safety standards are improving public health, but are also widening the gap between small livestock keepers and large commercial producers. The "livestock ladder" – by which smallholders climb up the scale of production and out of poverty – is missing several rungs.

Case studies show that small commercial livestock producers can be competitive, even in a rapidly changing sector, if they have appropriate institutional support and the opportunity cost of their labour remains low. Historical experience from member countries of the Organisation for Economic Co-operation and Development (OECD) shows that policy support in the form of subsidies and trade protection is very costly and has limited success in preventing the exit of smallholders from livestock production. Policy interventions aimed at improving smallholder productivity, reducing transaction costs and overcoming technical market barriers can be very helpful, but direct subsidies and protection could be counterproductive.

As economies grow and employment opportunities increase, the concomitant rising opportunity costs for labour often induce smallholders to abandon livestock keeping in favour of more-productive, less-onerous work in other sectors. This is an integral part of the economic development process and should not be viewed as a negative trend. Concerns arise when the pace of change in the livestock sector exceeds the capacity of the rest of the economy to provide alternative employment opportunities. Appropriate policy responses in this situation involve measures to ease the transition out of the sector, including the provision of social safety nets, and broader rural development policies, such as investments in education, infrastructure and growth-oriented institutional reforms. Smallholder agriculture should be the starting point for development, not the end-point.

Some livestock keepers are simply too poor, and their operations too small, to be able to overcome the economic and technical barriers that prevent their expansion into commercial production. Women typically face greater challenges than men, as they have poorer access to and control over livestock and other resources such as land, credit, labour, technology and services necessary to take advantage of growth opportunities. Most of the very poor depend on livestock as a safety net rather than using them as the basis of a commercial enterprise. Better access to animal-health services and a greater voice in livestock disease-control measures would improve their situation in the short run, but they would also benefit more from the creation of alternative social safety nets that protect livelihoods from external shocks. The vulnerabilities and constraints facing the poorest livestock keepers, and the important safety-net function livestock play for them, should be borne in mind. Indeed, the multiple roles of livestock in the livelihoods of people living in poverty should be considered in any policy decisions that affect them.

The agriculture sector is the world's largest user and steward of natural resources and, like any productive activity, livestock production exacts an environmental cost. The livestock sector is also often associated with policy distortions and market failures, and therefore places burdens on the environment that are often out of proportion to its economic importance. For example, livestock contribute less than 2 percent of global gross domestic product (GDP) but produce 18 percent of global greenhouse gas (GHG) emissions; it should be noted, however, that GDP underestimates the economic and social contribution of livestock as it does not capture the value of the numerous multifunctional contributions of livestock to livelihoods. There is thus an urgent need to improve the resource-use efficiency of livestock production and to reduce the negative environmental externalities produced by the sector.

Livestock grazing occupies 26 percent of the earth's ice-free land surface, and the production of livestock feed uses 33 percent of agricultural cropland. The expansion of land used for livestock development can contribute to deforestation in some countries, while intensification of livestock production can cause overgrazing in others. The increasing geographic concentration of livestock production means that the manure produced by animals often exceeds the absorptive capacity of the local area. Manure thus becomes a waste product rather than being the valuable resource it is in less-concentrated, mixed production systems. These wastes can become valuable resources again if proper incentives, regulations and technology, such as anaerobic digestion, are applied.

The concentration of animal production in close proximity to human population centres poses increasing risks for human health arising from livestock diseases. Livestock diseases have always interacted with human populations. Most strains of influenza, for example, are believed to have originated in animals. Furthermore, livestock pathogens have always posed a production challenge because, at the biological level, they compete with humans for the productive output of animals. Livestock diseases impose a heavy burden on the poor because poor livestock keepers live in closer proximity to their animals, they have less access to veterinary services, and the measures used to control certain disease outbreaks can threaten the basis of their livelihoods and the safety net they rely on in emergencies. Improving the management of livestock with a view to controlling diseases can provide significant economic, social and human-health benefits for poor people and society more broadly. This may require relocating livestock production away from human population centres in order to minimize the risk of disease transmission.

Transformations of the Livestock Sector

The State of Food and Agriculture last provided a comprehensive review of the livestock sector in 1982. Since then, the livestock sector has developed and changed rapidly in response to shifts in the global economy, rising incomes in many developing countries and changing societal expectations. The sector is increasingly expected to provide safe and plentiful food for growing urban populations as well as public goods related to poverty reduction and food security, environmental sustainability and public health. These trends and the challenges they entail were identified a decade ago by Delgado *et al.*, who coined the term the "livestock revolution" to describe the process that is transforming the sector:

> A revolution is taking place in global agriculture that has profound implications for human health, livelihoods, and the environment. Population growth, urbanization, and income growth in developing countries are fueling a massive increase in demand for food of animal origin. These changes in the diets of billions of people could significantly improve the well-being of many rural poor. Governments and industry must prepare for this continuing revolution with long-run policies and investments that will satisfy consumer demand, improve nutrition, direct income growth opportunities to those who need them most, and alleviate environmental and public health stress.

Rapid income growth and urbanization over the past three decades, combined with underlying population growth, are driving growth in demand for meat and other animal products in many developing countries. Supply-side factors, such as the globalization of supply chains for feed, genetic stock and other technology, are further transforming the structure of the sector. The sector is complex and differs according to location and species. A growing divide is emerging; large-scale industrial producers serve dynamic growing markets whereas traditional pastoralists and smallholders, while often continuing to support local livelihoods and provide food security, risk marginalization.

In many parts of the world, the transformation of the livestock sector is occurring in the absence of strong governance, resulting in market failures related to natural-resource use and public health. Interventions to correct market failures have been largely absent; in some cases, government actions have created market distortions. While the livestock sector is not alone in this regard, institutional and policy failures have led to opportunities presented by growth in the livestock sector being missed. As a result, the sector has not contributed as much as it might have to poverty alleviation and food security. Nor has growth in the sector been adequately managed to deal with the increasing pressures on natural resources or to provide control and management of animal disease. Correcting market failures is thus an important underlying rationale for public policy intervention.

Meeting Society's Expectations

The livestock sector, like much of agriculture, plays a complex economic, social and environmental role. Society expects the sector to continue to meet rising world demand for animal products cheaply, quickly and safely. It must do so in an environmentally sustainable way, while managing the incidence and consequences of animal diseases and providing opportunities for rural development, poverty reduction and food security. Given the large number of people who depend on livestock for their food security and livelihoods and the high environmental and human-health costs often associated with the sector, the challenge for policy-makers is to strike a fine balance among competing goals.

The livestock sector is one among many human activities contributing to the increasing pressure on ecosystems and natural resources: land, air, water and biodiversity. At the same time, the sector is increasingly constrained by this pressure on natural resources and the growing competition with other sectors for resources. There is also increasing awareness that climate change is creating a new set of conditions in which the sector must operate as well as imposing additional constraints on it.

Climate change will alter what men and women do, exposing them to different risks and opportunities. For example, men may migrate for work while women and youth will take on new responsibilities. Women tend to be more vulnerable to external shocks owing to unequal access to resources, lower level of education, increased work burden and poorer health.

Growing international trade in livestock and livestock products and the increasing concentration of livestock production in close proximity to large human populations have increased the risks of animal disease outbreaks and the emergence of new animal-related human-health threats. At the same time, inadequate access to veterinary services jeopardizes the livelihoods and development prospects of many livestock holders throughout the developing world.

Livestock can provide a pathway out of poverty for some smallholders, and policy-makers need to consider the different roles that livestock play in supporting livelihoods. For those smallholders who have the potential to compete as commercial enterprises, judicious policy and institutional support is needed to help them access technology, information and markets to improve their productivity. At the same time, the forces of economic change mean that some smallholders will need assistance to make the transition out of the sector. For others, especially the very poor, livestock primarily provide a safety-net function. The livestock sector requires renewed attention and investments from the agricultural research and development community and robust institutional and governance mechanisms that reflect the diversity within the sector. The livestock sector can contribute more effectively to improving food security and reducing poverty, but policy measures are required to ensure that it does so in ways that are environmentally sustainable and safe for human health.

This edition of *The State of Food and Agriculture* argues that the livestock sector could contribute more positively to society's goals, but significant policy and institutional changes are required. The rapid growth of the sector, in a setting of weak institutions and governance, has given rise to systemic risks that may have serious implications for livelihoods, human and animal health and the environment. Investments are required to improve livestock productivity and resource-use efficiency, both to meet growing consumer demand and to mitigate environmental and health concerns. Policies, institutions and technologies must consider the particular needs of poor smallholders, especially during times of crisis and change.

Current Trends in Livestock Sector

Rapid growth and technological innovation have led to profound structural changes in the livestock sector, including: a move from smallholder mixed farms towards large-scale specialized industrial production systems; a shift in the geographic locus of demand and supply to the developing world; and an increasing emphasis on global sourcing and marketing. These changes have implications for the ability of the livestock sector to expand production sustainably in ways that promote food security, poverty reduction and public health.

Trends in Consumption

Consumption of livestock products has increased rapidly in developing countries over the past decades, particularly from the 1980s onwards. Growth in consumption of livestock products per capita has markedly outpaced growth in consumption of other major food commodity groups. Since the early 1960s, consumption of milk per capita in the developing countries has almost doubled, meat consumption more than tripled and egg consumption increased by a factor of five.

This has translated into considerable growth in global per capita intake of energy derived from livestock products, but with significant regional differences. Consumption has increased in all regions except sub-Saharan Africa. Also, the former centrally planned economies of Eastern Europe and Central Asia saw major declines around 1990. The greatest increases have occurred in East and Southeast Asia and in Latin America and the Caribbean. The most substantial growth in per capita consumption of livestock products has occurred in East and Southeast Asia. China, in particular, has seen per capita consumption of meat quadruple, consumption of milk increase tenfold, and egg consumption increase eightfold. Per capita consumption of livestock products in the rest of East and Southeast Asia has also grown significantly, particularly in the Democratic People's Republic of Korea, Malaysia and Viet Nam.

Brazil too has experienced a rapid expansion in the consumption of livestock products – per capita consumption of meat has almost doubled, while that of milk has increased by 40 percent. In the rest of Latin America and the Caribbean, increases in consumption have been more modest, with some exceptions. The Near East and North Africa has seen a 50 percent increase in consumption of meat and a 70 percent increase in egg consumption, although milk consumption has declined slightly. In South Asia, including India, per capita consumption of livestock products has grown steadily, although meat consumption remains low. Among the developing-country regions, only sub-Saharan Africa has seen a modest decline in per capita consumption of both meat and milk.

In the developed countries overall, growth in per capita consumption of livestock products has been much more modest. The former centrally planned economies of Eastern Europe and Central Asia suffered a sudden drop in per capita consumption of livestock products in the early 1990s and consumption has not recovered since – as a result, per capita meat consumption in 2005 was 20 percent below its 1980 level.

Consumption of livestock products per capita in developing regions is still substantially lower than in the developed world, even though some rapidly developing countries are narrowing the gap. There is significant potential for increasing per capita consumption of livestock products in many developing countries. The extent to which this potential will translate into increasing demand depends on future income growth and its distribution among countries and regions. Rising incomes are more likely to generate additional demand for livestock products in low-income countries than in middle- and high-income countries.

Drivers of Consumption Growth

The growing demand for livestock products in a number of developing countries has been driven by economic growth, rising per capita incomes and urbanization. In recent decades, the global economy has experienced an unparalleled expansion, with per capita incomes rising rapidly.

Demographic factors also underlie changing consumption patterns of livestock products. An important factor has been urbanization. The share of total population living in urban areas is larger in the developed countries than in developing countries (73 percent compared with an average of 42 percent). However, urbanization is increasing faster in developing countries than in developed countries. In the period 1980–2003, the urban population in developing countries grew at average annual rates ranging from 4.9 percent in sub-Saharan Africa to 2.6 percent in Latin America, compared with an average of only 0.8 percent in developed countries.

Urbanization alters patterns of food consumption, which may influence demand for livestock products. People in cities typically consume more food away from home and larger amounts of pre-cooked, fast and convenience foods than do people in rural areas. Urbanization influences the position and the shape of consumption functions – the relationship between income and consumption – for food products. Estimating consumption functions for total animal-derived products in a sample of East Asian economies, Rae found urbanization to have a significant effect on the consumption of animal products, independently of income levels. Another implication of urbanization in many parts of the world is the growing concentration of animals in cities, in close proximity to humans, as people tend to move livestock activities to urban areas.

Social and cultural factors and natural-resource endowments can also significantly influence local demand and shape future demand trends. For example, Brazil and Thailand have similar levels of income per capita and urbanization, but livestock product consumption is roughly twice as high in Brazil as in Thailand. The influence of natural-resource endowments can be seen in the case of Japan, which has considerably lower levels of consumption of livestock products than other countries with comparable income levels, but compensates with higher levels of fish consumption. Natural-resource endowment affects the relative costs of different food commodities. Access to marine resources favours consumption of fish while access to natural resources for livestock production favours consumption of livestock products. Cultural reasons further influence consumption habits. In South Asia, for example, consumption of meat per capita is lower than income alone would seem to explain.

Production Trends

Developing countries have responded to growing demand for livestock products by rapidly increasing production. Between 1961 and 2007, the greatest growth in meat production occurred in East and Southeast Asia, followed by Latin America and the Caribbean. Most of the expansion in egg production was in East and Southeast Asia, while South Asia dominated milk production.

By 2007, developing countries had overtaken developed countries in terms of production of meat and eggs and were closing the gap for milk production. Trends in production growth largely mirror those for consumption. China and Brazil show the greatest growth, especially for meat. Between 1980 and 2007, China increased its production of meat more than sixfold; today, it accounts for nearly 50 percent of meat production in developing countries and 31 percent of world production. Brazil expanded meat production by a factor of almost four and now contributes 11 percent of developing-country meat production and 7 percent of global production.

In the remaining parts of the developing world, growth in meat output – as well as production levels – was lower, with the highest growth rates being in the rest of East and Southeast Asia and the Near East and North Africa. In spite of more than doubling meat production between 1980 and 2007, India's overall meat production levels remain low in a global context. However, after more than tripling milk production between 1980 and 2007, India now produces some 15 percent of the world's milk. Production of meat, milk and eggs also increased in sub-Saharan Africa but more slowly than in other regions.

Most of the increase in meat production has been from monogastrics; poultry meat production has been the fastest-growing subsector, followed by pig meat production. Increases from large and small ruminants have been much more modest. The result has been major changes in the composition of meat output globally, with significant differences between regions and countries.

Pig meat accounts for over 40 percent of global meat supplies, in part because of high levels of production and rapid growth in China, where more than half of world production takes place. The expansion of poultry meat production, which in 2007 accounted for 26 percent of global meat supplies, has been more widely distributed among both developed and developing countries, but again with China experiencing very high rates of growth. Globally, cattle production has increased much less and only in the developing countries. China and Brazil, in particular, have expanded production considerably and are each now responsible for around 12–13 percent of global cattle meat production. Meat from small ruminants remains of minor importance at the global level, but accounts for a significant portion of meat produced in the Near East and North Africa, sub-Saharan Africa and South Asia.

Drivers of Production Growth

Supply-side factors have enabled expansion in livestock production. Cheap inputs, technological change and scale efficiency gains in recent decades have resulted in declining prices for livestock products. This has improved access to animal-based foods even for those consumers whose incomes have not risen. Favourable long-run trends in input prices (e.g. feedgrain and fuel) have played an important role. Declining grain prices have contributed to increased use of grains as feed and downward trends in transportation costs have facilitated the movement not only of livestock products but also of feed. Recent increases in grain and energy prices may signal the end of the era of cheap inputs.

Increases in livestock production occur in two ways, or in a combination of the two:

- an increase in the number of animals slaughtered (in the case of meat) or producing (in the case of milk and eggs);
- increased output per animal (or yield).

Between 1980 and 2007, livestock numbers generally increased faster than yields. However, there are differences across regions and species.

Change in yield per animal is an important productivity indicator but it provides only a partial measure of productivity increases. It does not account for gains in terms of the rate at which animals grow and gain weight or any improved efficiency in input use or production factors. Other productivity indicators, although still imperfect, may contribute to providing a more complete picture of trends in livestock productivity.

Impacts of Technological Innovations

Technological change is the single most important factor in expanding supply of cheap livestock products. At the same time, it has affected the structure of the sector in many parts of the world.

Technological change refers to developments and innovations in all aspects of livestock production from breeding, feeding and housing to disease control, processing, transportation and marketing. Technological change in the livestock sector has mostly been the result of private research and development efforts aimed at commercial producers, in contrast with the publicly funded efforts aimed specifically at developing technological innovations that could be applied by smallholders that led to the green revolution in wheat and rice. As a result, technological innovations in the livestock sector have been relatively less widely available and applicable to smallholders. Little emphasis has been given to research on the public goods aspects of technology development for livestock, such as impacts on poor people or externalities related to the environment or public health.

The application of advanced breeding and feeding technology has spurred significant productivity growth, especially in broiler and egg production and the pork and dairy sectors. Technological advances, and thus productivity growth, have been less pronounced for beef and meat from small ruminants. The use of hybridization and artificial insemination has accelerated the process of genetic improvement. The speed and precision with which breeding goals can be achieved has increased considerably over recent decades. Genetic advances are much faster in short-cycle animals, such as poultry and pigs, than in species with a longer generation interval, such as cattle. In all species, feed conversion and related parameters, such as growth rate, milk yield and reproductive efficiency, have been major targets for breeding efforts, while features corresponding to consumer demands, such as fat content, are increasing in importance. While impressive advances have been made in breeds developed for temperate regions, results have been limited in development of breeds of dairy cows, pigs and poultry that perform well in tropical low-input environments.

Improvements in feed technology include balanced feeding, precision feeding, optimal addition of amino acids and mineral micronutrients, and development of improved pasture species and animal husbandry systems such as zero-grazing.

Animal-health improvements, including the increasing use of vaccines and antibiotics, have also contributed to raising productivity. These technologies have spread widely in recent years in a number of developing countries, particularly in industrial production systems close to major consumption centres.

Technological innovations in processing, transportation, distribution and marketing of livestock products have also significantly altered the way food is delivered to consumers (cold chains, longer shelf-life, etc.).

Growth in Livestock Trade

Growth in livestock trade has been facilitated by increasing consumption of livestock products and economic liberalization. Developments in transportation, such as long-distance cold-chain shipments (refrigerated transport) and large-scale and faster shipments, have made it possible to trade and transport animals, products and feedstuffs over long distances. This has allowed production to move away from the loci of both consumption and production of feed resources. Increasing trade flows also have implications for the management of animal diseases and a number of food-safety issues. Livestock products represent a growing proportion of agricultural exports. Their share of agricultural export value globally rose from 11 percent to 17 percent between 1961 and 2006. However, trade in crops – including feed crops – still dwarfs that of livestock products.

Between 1980 and 2006, the volume of total meat exports increased more than threefold. Exports of dairy more than doubled and exports of eggs almost doubled. The share of production entering international trade increased, except for sheep meat and eggs, reflecting the sector's increasing degree of openness to trade. The degree of trade openness has been particularly high for monogastrics.

Although the bulk of livestock produce is consumed within the country of production and does not enter international trade, livestock exports are important for a few countries. Since mid-2002, developing countries as a whole have been net exporters of meat. However, this masks large disparities between countries. Developing-country meat exports are dominated by the contribution of Brazil, the world's largest meat exporter. If exports from Brazil, China, India and Thailand are excluded, all developing regions are net importers of meat. Thailand has emerged as a major force in the global market for poultry, with net exports of almost half a million tonnes in 2006. All developing regions are increasingly dependent on imports of dairy products.

Brazil's performance in export for livestock products is particularly noteworthy. Over the last decade, the country has increased the quantity of poultry meat exports fivefold, and exports of pig and bovine meat have risen by a factor of 8 and 10, respectively. In nominal value, Brazil's net exports of livestock products went from US$435 million in 1995 to US$7 280 million in 2006. In 2006, Brazil's net exports accounted for 6 percent of global exports of pig meat, 20 percent of bovine meat and 28 percent of poultry meat. Brazil has increasingly taken advantage of low feed production costs for its livestock industry and is poised to remain an important producer of feedstuffs. The combination of abundant land and recent infrastructure developments has turned previously remote areas, such as Mato Grosso and the Cerrado region of central Brazil, into feed baskets. These two regions have the lowest production costs for maize and soybeans anywhere in the world. Since the early 1990s, Brazilian producers have actively taken strategic advantage of their position and have started to convert their feed into exportable surpluses of livestock commodities.

A particular source of concern is the net trade position in livestock products of the least-developed countries (LDCs). These countries are increasingly dependent on imports of livestock products – indeed, food commodities in general – to meet growing demand. The proportion of consumption met by imports has increased rapidly since 1996. As part of wider efforts to boost agricultural growth, expanding domestic supply could potentially contribute to economic growth, rural development and an improved external trade position.

Outlook for Consumption, Production and Trade

The factors that have encouraged growth in demand in developing countries – rising incomes, population growth and urbanization – will continue to be important over the coming decades, although the effects of some may weaken. Population growth, although slowing, will continue. While projections of the future growth in the world's population vary, a recent estimate suggests the world's population will exceed 9 billion in 2050.

Trends towards increasing urbanization are considered unstoppable. By the end of 2008, it is believed that, for the first time, more than half the world's population was living in towns and cities. By 2050, around seven out of every ten people are expected to be urban dwellers; by then, there will be 600 million fewer rural residents than now.

Income growth is generally considered to be the strongest driver of increased consumption of livestock products. Although short-term prospects are poor, with the global economy in a severe recession, medium-term prospects do suggest a recovery, albeit slow. In April 2009, the International Monetary Fund (IMF) projected a decline in global GDP of 1.3 percent in 2009, followed by growth of 1.9 percent in 2010, rising to 4.8 percent by 2014. According to the IMF, the exceptional uncertainty of the growth outlook means that the transition period will be characterized by slower growth than seen in the recent past.

The effect of economic growth on demand for livestock products depends on the rate of growth and where it occurs. Demand for livestock products is more responsive to income growth in low-income countries than in higher-income countries. Increasing saturation in per capita consumption in countries that have reached relatively high levels of consumption, notably Brazil and China, could lead to some slowing in demand. An important question is whether other major developing countries with low current levels of meat consumption will emerge as new growth poles, thus sustaining large increases in global demand. India, with its large population and low levels of per capita consumption of livestock products, has the potential to be a major source of new demand. However, opinions differ on the likely future contribution of India to global demand for livestock products.

A further question is to what extent continuing high food prices will dampen consumer demand, as consumers across the globe alter their eating habits. While it is difficult to forecast future feed and food price trends accurately, most analysts and observers agree that in the short to medium term, prices will remain higher than in the recent past, but that increased volatility of prices will become the norm.

Overall, the potential for expanding per capita consumption of livestock products remains vast in large parts of the developing world as rising incomes translate into growing purchasing power.

All indications are for continued growth in global demand for livestock products. In 2007, the "IMPACT" model developed by the International Food Policy Research Institute (IFPRI) projected an increase in global per capita demand for meat ranging from 6 to 23 kg, according to the region, under a "business-as-usual scenario" . The bulk of the increase is projected to be in developing countries. The largest numerical increases are projected for Latin America and the Caribbean and the East and South Asia and the Pacific regions, but a doubling – albeit from a low level – is foreseen for sub-Saharan Africa.

The model projects that growing demand will lead to increasing livestock populations, with the global population of cattle increasing from 1.5 billion to 2.6 billion and that of goats and sheep from 1.7 billion to 2.7 billion between 2000 and 2050. Demand for coarse grains for animal feed is also projected to increase over the period by 553 million tonnes, corresponding to approximately half of the total increase in demand.

The *OECD–FAO Agricultural Outlook, 2009–2018* presents projections for the coming decade. Although methodological and measurement differences between the two prevent direct comparison of precise figures, the OECD–FAO projections nevertheless confirm the trends indicated by the longer-term IFPRI projections. In spite of low economic growth in the first part of the projection period, OECD–FAO expect demand to continue growing, especially in the developing countries, driven by increasing purchasing power, population growth and urbanization. However, global meat consumption is expected to expand by an overall 19 percent compared with the base period, a slightly lower rate than over the previous decade (22 percent). Most of the increase is projected to occur in developing countries, with meat intake growing by 28 percent, compared with 10 percent at most in the developed and OECD countries. The increase is explained in part by population growth, but mostly reflects an increase in per capita consumption in developing countries of 14 percent – from 24 kg per person per year to more than 27 kg per person per year. Per capita consumption in developed countries is projected to increase by only 7 percent, from 65 kg to 69 kg. The smallest increase, of only around 3.5 percent, is projected for the OECD countries. Globally, demand for poultry is expected to continue to show the strongest growth.

According to the OECD–FAO projections from 2009 to 2018, 87 percent of global growth in meat production will occur outside the OECD area. For the developing countries, an overall increase in meat production of 32 percent is foreseen over the projection period.

The OECD–FAO projections for dairy suggest that demand, both per capita and overall, will continue to grow. The most rapid growth will occur in developing countries, where per capita demand is expected to increase at an annual rate of 1.2 percent. Overall production growth is projected at 1.7 percent per year from 2009 to 2018, with much of the increase coming from developing countries.

Feed demand is also projected to continue increasing. Use of coarse grains as feed is expected to grow by 1.2 percent a year. The total increase will amount to 79 million tonnes, to a total of 716 million tonnes, with most of the increase in developing countries. The projection excludes distiller dried grain (DDG), a by-product from ethanol production . Higher feed prices may lead to slower demand growth in the developing countries. Use of wheat as feed is also projected to increase slightly. Demand for oilseed meal is projected to grow by an annual rate of 3.8 percent in the non-OECD countries and 0.7 percent in the OECD countries. This, however, is only half the rate of growth seen in the previous decade.

Livestock Sector Diversity

The rapid growth of the livestock sector and projections for continued expansion are affecting the sector's structure. The livestock sector is characterized by large variations in the scale and intensity of production and in the nature and degree of linkages with the broader agricultural and rural economy. Further variation is found by species, location, agro-ecological conditions, technology and level of economic development. No single classification system can capture all of this diversity.

Looser terms such as "modern" and "traditional" are also used in this report to distinguish between parts of the livestock sector that have undergone greater or lesser degrees of economic and technological transformation in recent decades. Industrial systems are generally described as modern, although some grazing and mixed systems also use modern techniques such as breed selection and herd management. These terms are used to facilitate a comparative discussion of the costs, benefits and trade-offs implied by different systems for food security and livelihoods, environmental sustainability and human health, not to suggest that one is preferable to the other.

Grazing Systems

Grazing systems cover the largest land area and are currently estimated to occupy some 26 percent of the earth's ice-free land surface.

- *Extensive grazing systems* cover most of the dry areas of the world that are marginal for crop production. Such areas tend to be sparsely populated and include, for example, the dry tropics and continental climates of southern Africa, central, eastern and western Asia, Australia and western North America.These systems are characterized by ruminants grazing mainly grasses and other herbaceous plants, often on communal or open-access areas and often in a mobile fashion. The main products of these systems include about 7 percent of global beef production, about 12 percent of sheep and goat meat production and 5 percent of global milk supply.
- *Intensive grazing systems* are found in temperate zones where high-quality grassland and fodder production can support larger numbers of animals.These areas tend to have medium to high human population density and include most of Europe, North America, South America, parts of Oceania and some parts of the humid tropics. These systems are characterized by cattle (dairy and beef) and are based mostly on individual landownership. They contribute about 17 percent of global beef and veal supply, about the same share of the sheep and goat meat supply and 7 percent of global milk supply as their main outputs.

Mixed Farming Systems

In mixed farming systems, cropping and livestock rearing are linked activities. Mixed farming systems are defined as those systems in which more than 10 percent of the dry matter fed to animals comes from crop by-products or stubble or where more than 10 percent of the total value of production comes from non-livestock farming activities.

- *Rainfed mixed farming systems* are found in temperate regions of Europe and the Americas and subhumid regions of tropical Africa and Latin America. They are characterized by individual ownership, often with more than one species of livestock. Globally, around 48 percent of global beef production, 53 percent of milk production and 33 percent of mutton production originates from this type of production system.
- *Irrigated mixed farming systems* prevail in East and South Asia, mostly in areas with high population density. They are an important contributor to most animal products, providing about one-third of the world's pork, mutton and milk and one-fifth of its beef.

Industrial Production Systems

Industrial systems are defined as those systems that purchase at least 90 percent of their feed from other enterprises. Such systems are mostly intensive and are often found near large urban centres. Industrial systems are common in Europe and North America and in parts of East and Southeast Asia, Latin America and the Near East. They often consist of a single species (beef cattle, pigs or poultry) fed on feed.. They contribute slightly more than two-thirds of global production of poultry meat, slightly less than two-thirds of egg production and more than half of world output of pork, but are less significant in terms of ruminant production. These systems are sometimes described as "landless" because the animals are physically separated from the land that supports them. However, about 33 percent of global agricultural cropland is used to produce animal feed, so the term "landless" is somewhat misleading.

Animal Husbandry and Food Security

The livestock sector is one of the fastest-growing segments of the agricultural economy, particularly in the developing world. As demand for meat and dairy products in the developing world continues to increase, questions arise as to how this demand will be met and by whom. Parts of the sector, particularly poultry and pig production, have followed a trend similar to that in developed countries, where large-scale production units dominate output. The expansion of such trends across the whole livestock sector will have major implications for poverty reduction and food security. To date, the transformation of the livestock sector has occurred largely in the absence of sector-specific policies; this gap needs to be addressed to ensure that the livestock sector contributes to equitable and sustainable development.

Despite rapid structural change in parts of the sector, smallholders still dominate production in many developing countries. Livestock can provide income, quality food, fuel, draught power, building material and fertilizer, thus contributing to household livelihood, food security and nutrition. Strong demand for animal-based foods and increasingly complex processing and marketing systems offer significant opportunities for growth and poverty reduction at every stage in the value chain. These new market opportunities and livelihood options face rapidly changing patterns of competition, consumer preferences and market standards; these may undermine the ability of smallholders to remain competitive. They should also be carefully managed to ensure that women and men have the same prospects in this rapidly changing sector. Policy reforms, institutional support and public and private investments are urgently needed (i) to assist those smallholders who can compete in the new markets, (ii) to ease the transition of those who will exit the sector, and (iii) to protect the crucial safety-net function performed by livestock for the most vulnerable households.

Productivity growth in agriculture is central to economic growth, poverty reduction and food security. Decades of economic research have confirmed that agricultural productivity growth has positive effects for the poor in three areas: lower food prices for consumers; higher incomes for producers; and growth multiplier effects through the rest of the economy as demand for other goods and services increases. Agricultural growth reduces poverty more strongly than growth in other sectors. Recent research suggests that livestock sector growth can also promote broader economic growth and that smallholders can contribute to this. However, serious questions and policy challenges must be addressed if the potential of the livestock sector to promote growth and reduce poverty is to be met in a sustainable way.

Livestock and Livelihoods

Livestock are central to the livelihoods of the poor. They form an integral part of mixed farming systems, where they help raise whole-farm productivity and provide a steady stream of food and revenues for households. However, livestock's role and contribution to livelihoods in developing countries extends well beyond what is produced for the market or for direct consumption.

Livestock play many other important roles, including: as a provider of employment to the farmer and family members; as a store of wealth ; as a form of insurance; contributing to gender equality by generating opportunities for women; recycling waste products and residues from cropping or agro-industries; improving the structure and fertility of soil; and controlling insects and weeds. Livestock residues can also serve as an energy source for cooking, contributing to food security. Livestock also have a cultural significance – livestock ownership may form the basis for the observation of religious custom or for establishing the status of the farmer. The non-tradable roles played by livestock commonly vary between different parts of a country, and almost certainly among countries. They are also likely to change over time as economic conditions of livestock owners evolve.

The number of poor people who depend on livestock for their livelihoods is not known with certainty, but the most commonly cited estimate is 987 million or about 70 percent of the world's 1.4 billion "extreme poor".

Livestock keeping is pervasive among all income brackets of rural households In about one-third of the countries in the sample, poorer households are more likely to be engaged in livestock activities than are wealthier households. While there is no clear relationship between income level and engagement in livestock activities, it is clear that, in all the countries, even the poorest households commonly keep livestock.

The extent to which livestock contribute to income varies across countries and income levels. The share of household income derived from livestock ranges from less than 5 percent for many households to over 45 percent for middle-income households in Malawi. Although there is no systematic pattern, in several instances poor people earn a larger share of their income from livestock than do the wealthier households.

While the majority of rural households in the RIGA sample keep livestock, the average livestock holdings tend to be small, ranging from 0.3 tropical livestock units (TLUs) in Malawi to 2.8 TLUs in Ecuador. Holdings tend to be smaller in the African and Asian countries and larger in the Latin American countries. Also, although the proportion of households keeping livestock does not seem to be clearly associated with income level, average holdings tend to increase with wealth in 8 out of the 14 countries.

The proportion of livestock production sold, in terms of value, differs widely among countries in the sample, but not among expenditure quintiles. There seems to be no clear relationship between income levels and the share of livestock production that is sold. In several cases, the share of livestock production sold is less for the lowest-expenditure quintiles than for higher-expenditure quintiles, indicating that livestock are kept more for own consumption by the less well-endowed households, while they are kept as a source of cash income by better-off households. However, the pattern is not similar across the countries, with several countries revealing differences. In all the countries considered, more men than women own livestock, and households headed by men have larger livestock holdings than households headed by women. This is particularly true in the case of large animals (cattle and buffalo). Inequality in livestock holdings is particularly acute in Bangladesh, Ghana, Madagascar and Nigeria, where male-headed households keep more than three times as many livestock as do female-headed households (Anriquez, forthcoming). However, in the case of small livestock, particularly poultry, women play a much larger role. A large percentage of poultry production in Asia takes place in backyards, and it is mostly women who own and take care of the poultry. In Indonesia, 3.5 percent of poultry production takes place in the industrial sector, whereas 64.3 percent occurs in backyards. Poultry production in backyards by women is also substantial in Cambodia, the Lao People's Democratic Republic and Viet Nam. In many other countries and regions, women own poultry, sometimes in numbers greater than do men, and, unlike with other livestock, have the right to dispose of the poultry they raise without consulting men. The fact that women are responsible for poultry production in these areas has implications also for programmes to combat avian influenza.

The evidence from the RIGA database is generally consistent with the earlier findings. For example, Delgado *et al.* studied 16 different countries to compare the dependence on income from livestock of "very poor" and "not so poor" households. They found that most poor rural households are dependent on livestock to some extent, but the "not so poor" are likely to be much more dependent on income from animals than are the "very poor". In contrast, Quisumbing *et al.* found that, in many instances, the poor earn a larger share of their income from livestock than do the wealthy because they can exploit common property resources for grazing, so keeping production costs low.

Livestock Production and Human Nutrition

Undernutrition remains a persistent problem in many developing countries. The latest FAO figures indicate that nearly one billion people in the world are undernourished. Food security exists when all people at all times have access to adequate levels of safe, nutritious food for an active and healthy life. The livestock sector is central to food security, not only for rural smallholders who rely directly on livestock for food, incomes and services, but also for urban consumers, who benefit from affordable high-quality animal-based food. Livestock play an important role in all four main dimensions of food security: availability, access, stability and utilization.

Availability refers to the physical availability of adequate levels of food in a particular location. Food is made available through home production, local markets or imports. *Access* refers to the ability of people to acquire food. Even if food is physically present in an area, it may not be accessible if prices are very high or people lack purchasing power. Backyard and extensive grazing systems that rely on waste products and land that cannot be cultivated contribute unambiguously to the availability of food. By making efficient use of resources, they provide abundant low-cost food, contributing to the availability of and access to food. This role will become increasingly important as demand for livestock products continues to grow in coming years. At the same time, rapid growth in demand for livestock products means that, as noted earlier, one-third of all cropland is now used to produce livestock feed. Other things being equal, this competition for land traditionally reserved for the cultivation of other crops puts upward pressure on prices of staple foods and may undermine people's access to food.

Most rural households, including the very poor, keep livestock. Livestock contribute directly to food availability and access for smallholders, often in complex ways. Smallholders sometimes consume their home production directly, but they often choose to sell high-value eggs or milk in order to buy lower-cost staple foods. The indirect role of livestock in supporting food security through income growth and poverty reduction is crucial to overall development efforts. When calculating the economic contribution of livestock to individual households, it is also essential to recognize that men and women typically face different livelihood opportunities and constraints in managing livestock. Selling livestock allows resource-poor families to earn more income, but this may not always translate into improved nutrition, depending on whether it is men or women who have control over the income generated. The extent to which nutrition is improved depends on whether increases in income create more diverse diets. In the long run, there is an established connection between income growth and improved nutrition. However, in the short run, policy interventions may be necessary to promote increased consumption of foods of animal origin in the diets of the poor.

Stability is the third dimension of food security. Livestock contribute to the stability of food security of rural households by serving as an asset, a store of value and a safety net. Livestock can be used as collateral for credit, sold for income or consumed directly in times of crisis, thus buffering external shocks to the household such as an injury or illness of productive family members. Livestock also provide draught power, fertilizer and pest control in mixed farming systems, contributing to total farm productivity and hence to food security.

The fourth dimension of food security – *utilization* – is particularly relevant in the case of livestock and animal-based foods. Research shows that livestock products are an excellent source of high-quality protein and essential micronutrients such as vitamin B and highly bioavailable trace elements such as iron and zinc. This "bioavailability" is particularly important for mothers and small children, who find it difficult to obtain adequate levels of micronutrients in a plant-based diet. Small quantities of animal-based foods can provide essential nutrients for maternal health and the physical and mental development of small children.

The impact of poor nutrition on child growth and mental development is well documented and includes stunted growth and increased risk of infectious disease morbidity and mortality. Over the long term, undernutrition impairs cognitive development and school performance. Undernutrition is morally unacceptable, but it also comes at a high economic price. It reduces work performance and productivity in adults, lowers human capital development and constrains the potential for economic growth of countries. Undernutrition can also make women, men and children more vulnerable to diseases such as malaria, tuberculosis and HIV/AIDS.

Foods of animal origin can provide high-quality protein and a variety of micronutrients that are difficult to obtain in adequate quantities from foods of plant origin alone. Although essential minerals such as iron and zinc are also present in cereal staples, they have lower bioavailability in plant-based foods owing to their form and the presence of inhibitors of absorption such as phytates; they are more readily bioavailable in foods of animal origin.

Six nutritive elements that can be lower in primarily vegetarian diets and that are provided by animal-based foods include vitamin A, vitamin B_{12}, riboflavin, calcium, iron and zinc. Health problems associated with inadequate intake of these nutrients include anaemia, poor growth, impaired vision and blindness, rickets, impaired cognitive performance and increased risk of infectious disease morbidity and mortality, especially in infants and children. Animal-origin foods are particularly rich sources of all six of these nutrients, and relatively small amounts of these foods, added to a plant-based diet, can substantially enhance nutritional adequacy.

The high nutrient density of animal foods has a further advantage in food-based interventions targeting vulnerable groups such as infants, children and people living with HIV/AIDS, who may have difficulty consuming the large volumes of food needed to meet their nutritional requirements.

Available evidence indicates that in the poorest countries, where micronutrient deficiencies are most common, a moderate intake of foods of animal origin will improve the nutritional adequacy of diets and improve health outcomes. The Nutrition Collaborative Research Support Program reported strong associations between the intake of foods of animal origin and better growth, cognitive function and physical activity in children, better pregnancy outcomes and reduced morbidity resulting from illness in three parallel longitudinal observational studies in disparate ecological and cultural parts of the world, i.e. Egypt, Kenya and Mexico. These associations remained positive even after controlling for factors such as socio-economic status, morbidity, parental literacy and nutritional status.

Better access to foods of animal origin through the promotion of livestock together with nutrition education can thus be considered a strategic intervention for avoiding the poverty–micronutrient– malnutrition trap. Reviews of livestock interventions and their role in nutrition improvement and poverty reduction, although limited, show that livestock can play an important role in human nutrition and health and in poverty reduction in developing countries. Such interventions should be gender-specific to ensure that they effectively target food-insecure and vulnerable groups.

While there are strong arguments for promoting livestock in developing countries to improve nutrition and health, it is important to recognize that excessive consumption of foods of animal origin may have adverse health effects, such as obesity and associated chronic diseases, including heart disease and diabetes. In a recent major review of the evidence on food, nutrition, physical activity and cancer undertaken by the World Cancer Research Fund and the American Institute for Cancer Research, the panel of international experts involved in the review judged the evidence that red meats and processed meats are causes of colorectal cancer as "convincing" (red meats referring to beef, pork, lamb and goat from domesticated animals). There was considered to be limited evidence that fish and foods containing vitamin D (found mostly in fortified foods and animal foods) decrease the risk of colorectal cancer. But the Panel judged that milk probably protects against colorectal cancer. The Panel also noted limited evidence suggesting that red meats and processed meats are causes of other cancers.

A "nutrition transition" is occurring in rapidly growing economies in the developing world. Rapid changes in diet and decreasing levels of physical activity are leading to one form of malnutrition (obesity) replacing another (undernutrition). Growing consumption of high-fat animal products is one of several contributing factors. Using data on Chinese adults, for example, Popkin and Du have shown linkages between increased fat intake from animal-origin foods and a change in disease patterns. Sometimes these dietary shifts occur so rapidly that the two forms of malnutrition coexist in the same population. This has been referred to as the "double burden of malnutrition". Globally, by 2000, roughly equal numbers of people were overweight and underweight. The World Health Organization (WHO) estimates that more than 1.6 billion people are overweight, a number that is projected to increase to 2.3 billion by 2015.

The costs for developing countries that have to face this double burden of malnutrition are large. The human and financial costs of prevention and treatment of obesity and non-communicable diseases are high and place huge strains on existing health care systems. In the European Union (EU), the cost of obesity to society has been estimated at about 1 percent of GDP . In China, the economic cost of diet-related chronic diseases has already surpassed that of undernutrition – a loss of more than 2 percent of GDP. In Latin America and the Caribbean, such costs have been estimated at 1 percent of GDP of the region.

Such diet-related concerns are often considered lifestyle choices over which governments have little control. Governments can and do attempt to influence consumption patterns, however, through education, incentives and broader agricultural and food policies. Pacific island countries, which have the highest obesity rates in the world, have taken drastic measures to address diet-related health concerns. The Government of Fiji, concerned about the high fat content of sheep meat (mutton flaps) and turkey tails and the health consequences of importing such products, imposed an import ban on mutton flaps and instituted a ban on the sale (whether imported or locally produced) of these high-fat foods. Following the lead of Fiji, the Government of Tonga imposed an outright ban on the importation of mutton flaps. In 2007, the Government of Samoa also banned the importation of turkey tail meat in support of measures aimed at curbing the rapidly expanding problem of obesity and diet-related non-communicable diseases.

Livestock and Poverty Alleviation

Livestock production remains largely unchanged in the poorest countries, where consumption and production of meat and milk have increased little, if at all, over recent decades. Livestock are kept under traditional management systems by poor, small-scale farmers, for whom they are an important safety net, providing both high-quality food and cash in times of need. Non-tradable livestock products and functions remain important in these systems. Livestock products are processed and marketed largely through informal systems. Nevertheless, even in the poorest countries, an emerging urban middle class has stimulated a fledgling, albeit small, formal market that supplies certified, processed and packaged products.

Wherever rural poverty persists and non-farm employment options are limited, small-scale mixed crop–livestock systems persist. Globally, it is estimated that 90 percent of milk and 70 percent of ruminant meat is produced in mixed systems, as are more than one-third of pig and poultry meat and eggs. In these mixed systems, livestock typically generate up to one-third of farm income. Mixed crop–livestock systems thus make important contributions to the livelihoods, incomes and food and nutritional security of the rural poor.

In poor countries with pastoralist populations, traditional herders support subsistence livelihoods and sell live animals through local markets. In some countries in the Horn of Africa and the Sahel, pastoralists also supply cattle, sheep, goats and camels to traders who export live animals to traditional trading partners, mostly in the Near East and the growing coastal urban centres in West Africa. However, increasingly stringent sanitary standards threaten this trade. Pastoralism is under threat worldwide as mobility and access to traditional grazing areas become ever more restricted through border controls and the expansion of cultivation or, especially in parts of Africa, conservation-oriented activities. In addition, climate change appears to be making arid and semi-arid areas even drier and extreme weather events, including drought and floods, more common. Traditional coping mechanisms tend to fail in these situations and pastoralists are abandoning livestock production, voluntarily or involuntarily, in increasing numbers.

With economic growth, non-farm employment opportunities increase, rural wages rise, supermarkets extend their reach beyond urban centres and demand for livestock products increases further. Small-scale livestock keepers start to leave the sector as their need to keep a few livestock diminishes and the attractiveness and viability of the enterprise decline. The average size of holding of poultry and pigs tends to increase, although dairy herds often remain small. Even in rapidly growing markets, production and marketing of milk may still be dominated by the informal sector. Vertically integrated operators become larger and increasingly dominant, and small-scale poultry farmers find it increasingly difficult to stay in business, although small-scale pig keepers tend to be more successful in this regard.

Expanding markets for livestock products would appear to offer opportunities for improving the incomes of the many rural poor who depend on livestock for their livelihoods. However, while the growth and transformation of the sector have created opportunities, the degree to which these can be harnessed by people living in poverty and in marginalized areas is not clear. The rapid changes in food demand in some parts of the developing world have required the livestock sector to produce as much as possible, as quickly as possible, as cheaply as possible and as safely as possible. This emphasis on speed, quantity, price and safety has created a bias towards large-scale intensive production, especially in some subsectors such as poultry and pigs.

The nature of the livestock sector has changed dramatically in some parts of the world, although the impacts vary among countries, species and genders. Countries where per capita consumption of livestock products has increased dramatically over recent decades, especially the rapidly emerging economies such as Brazil, China and India, are diverging from those where consumption remains static or is decreasing, such as much of sub-Saharan Africa. At the same time, within the countries in which transformation of the livestock sector has taken off, a widening gulf is opening between a small-scale traditional sector, where women play an active role, at one extreme and a growing large-scale, intensive sector, in which men tend to dominate, at the other.

As economic growth continues to drive livestock development, there is increasing pressure for parts of the sector to industrialize. Overall, while strong growth within the sector should be seen as a positive sign of economic development, the *speed* of change may put pressure on smallholders. Some livestock producers will probably find it hard to adjust quickly enough to safeguard their income and, in some cases, their food security. Experiences in OECD countries from the 1950s onwards show that changing production structures require labour markets to adjust. However, when the transition is extremely rapid, as is happening in the livestock sector in many places today, the implications for poverty and food security can be dramatic and warrant intervention.

For the past decade, researchers and policy-makers have assumed that growth in the livestock sector was primarily demand-driven and that policies should aim at supporting demand growth and improving market opportunities . Recent research however, shows that supply-side factors are also important. In many developing countries, growth in the livestock sector actually leads to GDP growth. This means that policies aimed directly at promoting productivity growth in the livestock sector can support broader economic growth. The complex value chains for animal-based foods – from feed and animal production through processing and marketing – mean that growth in the sector can generate strong backward and forward economic linkages and employment opportunities, with important impacts on growth that favours the poor. Creating the conditions necessary for smallholders to take advantage of these opportunities is a major policy challenge, requiring careful attention also to gender issues and environmental dimensions. Overcoming supply constraints for smallholders and increasing their productivity are important both to allow them to benefit from the demand-led gains and to allow the sector to play its role as a driver of growth.

Demand growth will continue to be a significant factor driving trends in the livestock sector in the future. However, supply-side factors, including relative competitiveness of different production systems and supply constraints faced by different producers, will also shape the sector and influence its contribution to poverty alleviation.

Reducing rural poverty through agricultural development alone is difficult. The challenge for livestock development is to foster development in rural areas in ways that benefit entire rural communities, and not only those who are engaged in livestock activities. Rural development policies can further facilitate the transformation of the sector by creating alternative opportunities for income generation and employment.

The objective of livestock sector development policies should be to enhance the competitiveness of smallholder production systems, where feasible, while mediating sector transition and protecting the poorest households, which rely on livestock as a safety net. Poor people need to be considered broadly, including their roles as consumers, market agents and employees, as well as small-scale producers and, possibly, as providers of environmental services. All of this needs to take into account gender-related issues to ensure that the needs, priorities and constraints of women and men, both young and old, are taken into consideration in the design and implementation of livestock sector development policies.

References

Humphrey, J. (1980). *The classification of world livestock systems.* A study prepared for the Animal Production and Health Division. FAO. AGA/MISC/80/3.

Jahnke, H.E. (1982). *Livestock production systems and livestock development in tropical Africa.* Kiel, Germany: Kieler Wissenschaftsverlag Vauk.

Ruthenberg, H. (1980). *Farming systems in the tropics.* Clarendon Press. Oxford.

Wilson, T. (1994). *Integrating livestock and crops for the sustainable use and development of tropical agricultural systems.* AGSP-FAO.

Winrock International. (1992). *Animal Agriculture in Developing Countries: Technology Dimensions.* August 1992. Morrilton, Arkansas, USA.

2

Animal Housing

The main purpose of livestock production is to convert the energy in feed into products that can be utilized by human beings, such as milk, eggs, meat, wool, hair, hides and skins, draught power and manure (fertilizer). Traditional, extensive livestock production involving indigenous breeds and low-cost feeding will usually have low performance and can therefore only justify minimal, if any, expenditure for housing. However, where improved breeds, management and feeding are available it will usually be economically beneficial to increase the production intensity.

Although this can be facilitated by, among other things, the construction of buildings and other livestock structures to provide for some environmental control, reduced waste of purchased feedstuffs and better control of diseases and parasites, this rule is not invariable. For example, it is difficult to identify an economic benefit in sheep production arising from the use of anything but the least expensive buildings. At the other end of the scale, a relatively expensive farrowing house, providing a high level of environmental control, may improve the survival rate in piglets sufficiently to justify the cost and add to the profitability of the production unit.

The planning and design of any structure for a livestock production system involves many alternatives for each of numerous variables and can therefore be turned into a complex and theoretical subject, but is usually far simpler in reality. However, every facet of the design, including the production system, equipment, building materials, layout and location, will play a part in determining the profitability of production and any variation in one of them may significantly affect the profitability of the whole.

One special difficulty when designing livestock structures for tropical climates is that, up to now, most research and development has been concerned with the conditions in temperate or cold climates. Any recommendations derived from such experiments and applied uncritically in warm climates may result in an adverse environment for the animals and in very high building and operating costs.

Animal Behaviour

A basic understanding of domestic animal behaviour and the relationship between human beings and farm animals can contribute greatly to increased economic benefit in animal husbandry and to easier handling of the animals. The importance of animal behaviour aspects in the design of animal housing facilities generally increases with the intensity of production and the degree of confinement. Many modern farming systems greatly reduce the freedom of animals to choose an environment in which they feel comfortable. Instead they are forced to resort to an environment created by humans.

Animals that can exercise their natural species-specific movements and behaviour patterns as far as possible are less likely to be stressed or injured and will therefore be more productive. However, in the practical design of an animal production system and any buildings involved, many other factors, such as feeding, management, thermal environment, construction and economics, can be equally or even more important. The animals can, to some extent, adapt their behaviour to suit a bad design and, on a long-term basis, they can be changed by breeding and selection, but generally it will be much easier to tailor the husbandry and building design to the animals.

The lifespan of a building is usually 5 to 15 years, which makes it clear that even a small increase in production or decrease in the frequency of injury and disease, feed waste or labour requirements for animal handling will repay all the thought and care that has been put into the design, layout and construction of the building. Furthermore, it may cost as much to construct a building that is poorly designed and equipped for the animals as one that works well.

Behaviour Patterns

Farm animals are born with certain fixed behaviour patterns, such as pecking in chickens and nursing in mammals, but most behaviour patterns develop through play and social contact with other animals of their kind and under the influence of environmental stimulation and genetic factors. Although behaviour variation within a species is caused mainly by differences in the environment and between the sexes, breed, strain and individual variance also have an influence. Domestic animals show great ability to modify their behaviour patterns in relation to environments and to learn from experience.

Animals often form a daily cycle of habits caused by the uniformity of husbandry; for example, the regular variation in light during night and day relate to internal physiological rhythms. This is why cows gather around the barn just before milking time. Some behaviour patterns change from season to season, partly in response to the changing weather. Cows tend to be more active during the night in the hot season and, if outside, spend less time lying down during the wet season. Many domestic animals show a slight seasonal breeding pattern. Domestic animals under conditions of close captivity frequently display abnormal behaviour such as stereotyped movements or inappropriate sexual behaviour, particularly if they are unable to escape from, or adapt to, the situation. However, many disturbed behaviours have more complex causes. For example, tail and ear biting in pigs may be associated with boredom, breakdown of social order, an excessively high stocking rate, insufficient fibre content in the feed, malnutrition, poor ventilation leading to high humidity and overheating, lack of bedding, inadequate trough space and watering points, skin disease, parasites, teething problems, etc.

Social Rank Order

Domestic animals are highly sociable and naturally form groups. Males and females form separate groups, except during the breeding season, and the young tend to form small groups in the proximity of the female group. When strange adult animals meet for the first time they are likely to fight to establish dominant/subordinate relationships. The resulting pecking order, or social rank order, in which one or two animals are invariably dominant is usually formed quickly. While physical age and weight are the main factors determining social rank, sex, height and breed can also be an influence. The group can live in relative harmony as long as each animal knows its place and gives way to animals of higher rank. However, the order is seldom strictly hierarchic or static. Some animals of low rank may dominate others whose positions are normally higher, and fast-growing and maturing animals may move up the ladder.

The introduction of new animals into a group or the mixing of groups will normally lead to fighting until a new social order is established, and this may cause a growth check as well as injury.

The normal response to aggressive behaviour in a group with an established social order is for the subordinate animal to move away. The building layout must allow space for this and therefore narrow passages and corners where one animal can be trapped by another should be avoided in pens and yards. The order usually remains stable provided the group is small to ensure that all animals in it can remember the positions of the others, i.e. fewer than 80 cows, between 12 and 15 pigs or about 100 chickens.

Animal Behaviour and Building Design

Animal behaviour can influence the design of structures, as demonstrated in the examples given below.

Cattle normally live in herds but, when giving birth, cows attempt to find a quiet, sheltered place away from the disturbance of other cows and humans. The cow needs to be alone with her calf for some time after birth for the cow/calf bond to be established. When a cow that is confined in a loose housing system is approaching calving, it should be removed from the herd and placed in an individual pen.

Hens spend considerable time in the selection of a nest, which is on the ground. Nesting is characterized by secrecy and careful concealment. Hens in deep-litter systems therefore, sometimes lay eggs on the floor instead of in the nest boxes, especially if the litter is quite deep or there are dark corners in the pen.

To avoid this, plenty of fresh litter is provided in the nests, and they are kept in semidarkness and fitted with a rail in front so that birds can inspect the nests prior to entry. An additional measure is to start with nest boxes on the floor and slowly raise them to the desired level over a period of days.

Sows are nest-builders and should be transferred to clean farrowing pens one to two weeks before giving birth, and given some bedding with which they can build a nest. Oestrus, especially in gilts, is increased by the smell, sight and physical presence of a boar. Gilts and sows awaiting mating should therefore be kept in pens adjoining the boar pen.

Cattle prefer to be able to see while drinking, therefore more animals can drink at once from a long, narrow trough than from a low round one. With cattle (and hens), feeding is typically a group activity, therefore space at the feed trough must be provided for all the animals at one time. At pasture, undersized feed or water troughs can result in inadequate feeding and watering of the animals that are lowest in rank because these animals could well be excluded from the trough but, despite this, they still tend to leave with the rest of the herd after feeding or watering.

To prevent wasting feed, a trough should be designed to suit the particular behaviour pattern that each species exhibits while feeding, i.e. pecking in hens; rooting with a forward and upward thrust in pigs; and wrapping their tongue around the feed (grass) and jerking the head forward in cattle.

Artificially reared calves bunt the bucket instead of the cow's udder, and this requires a sturdy holder for the bucket. The habit of suckling each other is a problem in dairy calves. The problem can be reduced by making the calves suckle harder and longer for their feed by using a rubber teat rather than a bucket and by giving them access to dry feed. Assuming intersuckling is not a problem, a group pen for calves is more natural than individual pens, and helps to ensure normal activity and resting.

Sheep are vigilant and tight-flocking, and respond to disturbances by fleeing. When designing handling facilities, these characteristics should be taken into account. A race should be straight, level, fairly wide, have no blind ends, and preferably have close-boarded sides. Sheep that are following should be able to see moving sheep ahead, but advancing sheep should not see the sheep behind as they will tend to stop and turn around.

Sheep move best from dark into light areas and dislike reflections, abrupt changes in light contrast and light shining through slats, grates or holes. Handling facilities should be examined from the sheep's eye level, rather than from human eye level to detect flaws in the design.

Animal Environmental Requirements

The capacity of an animal to produce differs between species, breeds and strains as a result of genetic factors. However, a complex set of interrelated animal husbandry factors will influence the animal's ability to utilize that capacity for growth, development and production.

Progress in breeding and feeding to further increase production and efficiency can be limited by environmental factors. Research into these factors has therefore been increasing in recent years, especially in countries with intensive animal production systems. Animal housing design is mainly concerned with the physical environment, in particular climatic and mechanical factors. However, all other factors should also be considered in order to create a good layout, where healthy, high yielding animals can be provided with correct feeding, can be easily handled and can produce without stress or suffering physical harm.

Heat Regulation

All domestic livestock are *homeotherms,* that is to say, they maintain a relatively constant internal body temperature, usually within a 1–2 °C range. The body temperature of most domestic animals is considerably higher than the environmental temperature to which they are exposed most of the time. They maintain their body temperature by balancing internal heat production and heat loss to the environment. The *hypothalamus gland* acts as a body thermostat by stimulating mechanisms to counteract either high or low ambient temperatures. For example, increased conversion of feed-to-heat energy is used to counteract low ambient temperatures, while increased respiration (rate and volume) and blood circulation in the skin counteracts high ambient temperatures.

Table 1: Normal body temperatures of domestic animals and humans temperature (°c)

Animals	*range*	*Average*
Dairy cow	38.6	38.0–39.3
beef cow	38.3	36.7–39.1
Pig	39.2	38.7–39.8
Sheep	39.1	38.3–39.9
Goat	39.5	38.7–40.7
Horse	37.9	37.2–38.2
Chicken	41.7	40.6–43.0
Human	37.0	—

Varying the temperature also results in changed behaviour. Most animals reduce their level of activity in a hot environment and, for example, pigs lie clustered in a heap at low temperatures, while they lie spread out with extended limbs at high temperatures. This would suggest an increased space requirement for pigs in a warm, tropical climate. The body can tolerate short periods of heat stress but if the ambient temperature exceeds the body temperature for an extended period, it may prove fatal.

When feed is converted by the animal's metabolism for the production of milk, eggs, meat, offspring, etc., heat is produced as a by-product. An increased production level (and hence feed requirement) will therefore result in increased internal heat production. High-yielding animals are consequently more likely to suffer from heat stress in a hot climate than low-yielding ones.

Feeding fibre-rich, low-digestible feedstuffs, such as hay, will result in high heat-production because of increased muscular activity in the alimentary tract and, in ruminants, increased micro-organism activity in the rumen. An increased share of concentrates in the feed may therefore reduce heat stress in an animal under hot climatic conditions.

Animal Moisture and Heat Production

Heat is produced centrally in the deep body. The surplus is conducted to the skin surface where it is transferred to the atmosphere as sensible heat by means of convection, conduction and radiation, and as latent heat through the evaporation of moisture from the lungs and skin. Increasing the ambient temperature, resulting in a smaller temperature difference between the body surface and the air, will decrease the amount of heat that can be emitted as sensible heat. Instead, a larger proportion is given off as latent heat, that is to say, heat employed to vaporize moisture.

The heat and moisture produced by the animals confined in a structure must be removed by ventilation. In the tropics, sufficient air flow is usually provided by the use of open-sided structures. However, if an enclosed building is used, a range of ventilation flow rates must be provided for in the building design. The minimum ventilation rate should remove the moisture produced, but retain as much sensible heat as possible during cold periods. The maximum ventilation rate should remove enough of the sensible heat produced so that a small temperature difference, usually 2–4 °C, can be maintained between inside and outside. It should be noted that ventilation alone can maintain the building at only slightly above ambient temperature.

Climatic Factors

Temperature

The overriding environmental factor affecting the physiological functions of domestic animals is temperature. For most farm animals, a mean daily temperature in the range 10–20 °C is referred to as the 'comfort zone'. In this range, the animal's heat exchange can be regulated solely by physical means, such as the constriction and dilation of blood vessels in the skin, ruffling up the fur or feathers and regulation of the evaporation from lungs and skin.

At the upper and lower critical temperatures, physical regulation will not be sufficient to maintain a constant body temperature and the animal must, in addition, decrease or increase its metabolic heat production A further decrease or increase in temperature will eventually bring the temperature to a point beyond which not even a change in heat production will be sufficient to maintain *homeothermy*.

A very young animal, lacking fully developed temperature-regulating mechanisms, particularly the ability to increase heat production by increased metabolism, is much more sensitive to its thermal environment and requires higher temperatures.

Humidity

Poultry do not have sweat glands, so all evaporative heat loss must originate from the respiratory tract. Other livestock species have varying abilities to sweat and, in descending order, they are as follows: horse, donkey, cattle, buffalo, goat, sheep and pig.

In a hot, dry climate evaporation is rapid but, in a hot humid climate, the ability of the air to absorb additional moisture is limited and inadequate cooling may result in heat stress.

Excessively low humidity in the air will cause irritation of the mucous membranes, while excessively high humidity may promote the growth of fungus infections. High humidity may also contribute to decay in structures. If possible, keep the relative humidity in the range of 40 percent to 80 percent.

Radiation

The heat load on a grazing animal can be increased considerably by direct solar radiation and radiation reflected from clouds or the ground. A white hair coat will absorb less radiant energy than a dark one, but the heat penetrates deeper into a white, loose coat. Air movements will dispel the heat and reduce the differences. Solar radiation may also adversely affect the animal's skin, in particular breeds with unpigmented skin.

Heat gain by radiation can be effectively reduced by the provision of a shaded area. It must, however, be sufficiently large to allow space between the animals to avoid reducing heat loss by other means. Grass-covered ground in the surroundings of the shade will reflect less radiation than bare soil.

Air Movements

Air movements assist heat loss by evaporation and by conduction/convection, as long as the air temperature is lower than the skin temperature. When the air temperature approaches skin temperature, rapid air movements are experienced as comfortable but, at low temperatures, they lead to excessive cooling of unprotected skin areas (cold draught). In addition, air movements are required to remove noxious and toxic gases and to supply the animal with fresh air for breathing. A wind velocity of 0.2 m/s is generally regarded as a minimum requirement, but it can be increased to 1.0 m/s when the temperature is nearing the upper critical level, or more when it rises beyond that.

Precipitation

Heavy rain may penetrate the fur of an animal and decrease its insulation value. In such circumstances, a strong wind can lead to excessive cooling. However, a naturally greasy hair coat will resist water penetration and with the provision of a shelter for the animals the problem may be avoided altogether.

Microbiological Environment

Disease remains a major profit-limiting factor in animal production in many tropical countries. Sanitary control measures should be incorporated into any building design, so that a good hygienic standard can be easily maintained. An animal that is well fed and watered, as well as being in good condition, will have high resistance to disease. Good management can do much to remove or reduce the effects of adverse environmental factors, such as climatic stress, which would otherwise weaken the body's natural defences.

Newborn stock should always receive colostrum (first milk), which contains antibodies. It takes time for an effective immune system to develop in an animal and therefore good hygiene is of special importance in facilities for young animals. Pens, in particular those for calving and farrowing, should be constructed of easily cleaned and disinfected materials and be free from corners and recesses where manure and dirt can accumulate.

The whole building should be cleaned and disinfected periodically, and any pen that is emptied should be thoroughly cleaned before other animals are transferred to it. Rearing and fattening of young animals should be organized so that the building can be emptied, cleaned and disinfected between batches. This 'all-in, all-out' policy is particularly beneficial for disease control, where the animals are bought from outside the farm, and in finishing units for pigs, as well as broiler and layer houses.

Diseases are transmitted in many ways, including direct contact between animals, airborne micro-organisms, biting insects and ticks, manure, soil, contaminated feed and water, birds and rodents, as well as the stockperson's boots. Direct contact between animals can be reduced by decreasing the number of animals in each group and by constructing solid partitions between pens. However, solid walls may obstruct air movement and thus contribute to heat stress. Ideally, the waste handling system should prevent animals of different groups from coming into contact with one another's manure. Young animals, in particular, must be protected from contact with manure from adult animals.

Good husbandry includes regular observation of the animals to detect any change in behaviour that could indicate disease. Sick animals should be separated from the herd immediately to prevent further spread of infectious disease and to allow them to rest. The sick animal should be isolated in a pen kept especially for this purpose, which should ideally be in a separate building.

Newly acquired animals, and animals returning from a market or any other place where they may have been exposed to the risk of infection, must be quarantined for an adequate length of time to detect any disease that they may be carrying before they are allowed into the herd.

Other Environmental Factors

As far as we know, *acoustic factors* have only a marginal effect on the animal's development and production. Nervous animals may, however, react adversely to intermittent sudden noises. Pig squeals prior to feeding can become a hazard to the stockperson's hearing. Soft radio music in a milking parlour may have a soothing effect on the cows.

Day length or photoperiod varies with latitude and season and has a direct influence on animal performance, especially on the breeding season for sheep and poultry egg production. Under natural conditions, there is a correlation between the length of day and the rate of laying. Artificial light is used in the temperate zone to equalize egg production throughout the year. Additional hours of light before dawn and after dusk are recommended in hot climates to encourage the hens to eat during the cooler hours.

Dust can carry micro-organisms, which may cause an outbreak of disease.

Toxic and noxious gases are produced by manure that accumulates in buildings or storage facilities, especially during the agitation of manure slurry stored in a pit in a building, when harmful amounts of gas can be released. However, problems with gases are not likely to arise in the open-sided buildings used in the tropics.

Cattle Housing

Cows play an extremely important role in most African cultures. The ownership of cattle will often be the deciding factor in a person's social position in the community because the herd may be the only practical way of accumulating wealth. However, of greater importance is the fact that cattle represent a source of high-protein food, in the form of both milk and meat.

This section focuses on housing requirements for cattle kept primarily for milk production. Little or no housing is required for herds maintained only for beef production, and special handling and support facilities are discussed separately.

Much of the dairy farming in east and southeast Africa occurs at elevations of 1 500 metres or more. European breeds have been successfully established under these circumstances. However, European bulls crossed with zebu cows have produced animals that are more tolerant of high temperatures than the European breeds and are significantly better producers than zebu.

Whether purebred or crossbred, they will not provide a profit to the farmer if they are left to find their own feed and water and are milked irregularly. Experience has shown that cattle respond favourably to good management, feeding and hygiene, all of which is possible in a system with suitable housing.

Herd Profiles

The composition and management of cattle herds varies considerably. At one extreme, nomadic herdsmen graze their entire herd as one unit.

Smallholders with only a few head may keep their heifer calves for replacements or sell them. Commercial dairy producers typically have about four-fifths of their cows milking and one-fifth waiting to calve, while heifers of 10 months to calving age, plus calves of various ages, will approximately equal the number of milkers. Mature dairy cows are bred annually and are milked for 300–330 days after calving. On closer examination, several factors will be found to influence the number of animals of various categories found in a dairy herd. In a herd of say, 24 cows, with calving evenly distributed throughout the year and a 12-month calving interval, an average of two calves will be born per month. The calves are normally kept in individual pens for two to three months. There is therefore a requirement for four to six pens in a herd of 24 cows. However, the need for calf pens is halved in herds where the bull calves are sold or otherwise removed from the herd at one to three weeks of age. A longer calving interval and high mortality among the calves will decrease the required number of calf pens, while concentration of the calving season in the herd will increase the pen requirements. If all calving is concentrated in six months of the year, the requirement for calf pens will be doubled.

A number of cows in a dairy herd will be culled each year because of low milk yield, infertility, disease, old age, etc. These cows are best replaced with young stock from their own herd, because any animals acquired from outside the farm may bring disease to the herd. Cows are commonly culled after three to five lactations, corresponding to a replacement rate of 20 percent to 30 percent per year.

In herds with very intensive production there is a tendency towards a higher replacement rate, but it cannot exceed 40 percent if the heifers are obtained exclusively from the herd itself. This is due to the fact that only about half of the calves born are female, and some of these will die or be culled before first calving as a result of disease, infertility or other factors.

The number of maturing heifers will increase in line with an increase in the age of heifers at first calving, a higher replacement percentage and a shorter calving interval. Concentrated calving may slightly increase the number of animals during some periods of the year, and will greatly affect the distribution of animals in the different age groups. The age at first calving of heifers of European breeds is typically 24–27 months, while heifers of the slower-maturing zebu cattle are often aged 36 months or more.

Maturing heifers require little or no housing facilities in the tropics. Knowledge of their exact number and distribution in various age groups during different months is therefore not as important to a building designer as to the manager of the herd.

Heifers should be introduced into the dairy herd at least a couple of months prior to their first calving, to enable them to learn and become adjusted to the handling and feeding routines. In loose housing systems with free stalls (cubicles), or in tie-barns, this may slightly increase the need for stalls, but normally the heifer will simply take over the stall used by the culled cow that it replaces.

In herds where cows are taken to a special calving pen during calving, one such pen per 30 cows is sufficient, because the cow and her calf will spend only a few days there. However, in herds where calving is concentrated in a short period, the requirement can increase to one calving pen per 20 cows. The pen should be at least 3.3 metres by 3.3 metres.

General Housing Requirements

As has already been pointed out, cattle will produce milk and reproduce more efficiently if they are protected from extreme heat, i.e. temperatures of 25–30 °C, and particularly from direct sunshine. Thus in tropical and subtropical climates, providing shade becomes an important factor.

If cattle are kept in a confined area, it should be free from mud and manure in order to reduce hoof infection to a minimum. Concrete floors or pavements are ideal where the area per cow is limited. However, where ample space is available, an earth yard, properly sloped for good drainage, is adequate.

Shade from The Sun

With these needs in mind, a shade structure allowing 2.5–3 square metres per animal will give the minimum desirable protection for cattle, whether for one animal belonging to a smallholder or many animals in a commercial herd. A 3 metre by 7 metre roof will provide adequate shade for up to eight cows. The roof should be a minimum of 3 metres high to allow air circulation.

If financially feasible, the entire area that will be shaded at some time during the day should be paved with good-quality concrete. The size of this paved area depends on the orientation of the shade structure. If the longitudinal axis is east to west, then part of the floor under the roof will be in shade all day. Extending the floor by approximately one-third of its length on the east and on the west results in a paved surface being provided for the shaded area at all times.

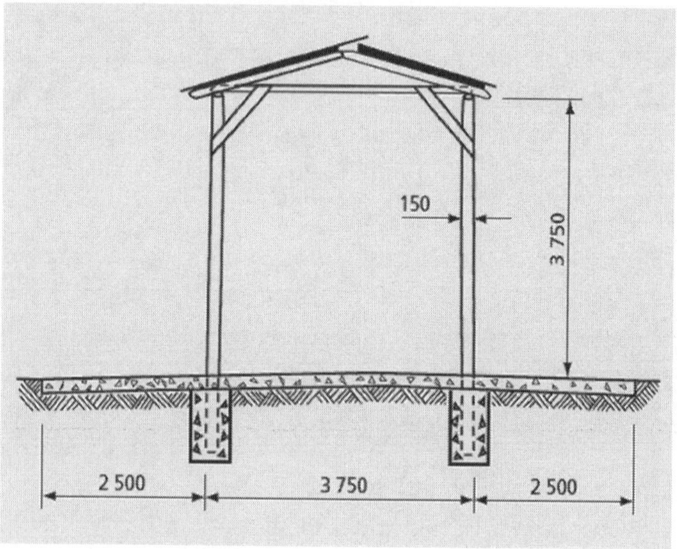

Fgure 1: Sunshade with insulated corrugated steel roof

If the longitudinal axis is north to south, the paved area must be three times the roof area i.e. one-third to the east, one-third to the west and one-third underneath. Obviously this means an increase in the cost of paving.

Yards

If space is severely limited and only 4–5 square metres per cow is available, then concrete paving is highly desirable. If 40–60 square metres per cow is available, then unpaved yards should be quite satisfactory, provided that the feed and shade areas are paved and the yard is graded for good drainage.

If the smallholder is unable to afford an improved structure, such as a shade or a paved area for feeding, then conditions can be prevented from becoming intolerable by building mounds of earth in the yard with drainage ditches between them. Between 20 square metres and 30 square metres per cow will keep the animals out of the worst of the mud. The soil in the mounds can be stabilized by working chopped straw, or straw and manure, into the surface. A number of trees in the yard will provide sufficient shade.

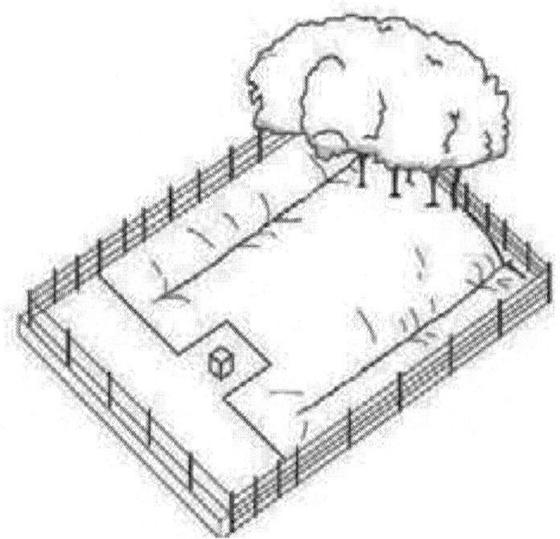

Figure 2. Yard with fence-line feed trough, paved feed area and earth mound

Deep-bedded Sheds

In a deep-bedded system, straw, sawdust, shavings or other bedding material is periodically placed in the resting area so that a mixture of bedding and manure builds up in a thick layer. Although this increases the bulk of manure, it may be easier to handle than wet manure alone. This system is most practical when bedding is plentiful and cheap.

Loose Housing with Free Stalls (cubicles)

Although simple yard and shade, or yard and bedded-shed, systems are entirely satisfactory in warm climates, particularly in semi-arid areas, some farmers may prefer a system with somewhat more protection. A loose housing yard and shed with free stalls will satisfy this need. Less bedding will be required and less manure will have to be removed. Free stalls must be of the right size in order to keep the animals clean and to reduce injuries to a minimum.When stalls are too small, injuries to teats will increase and the cows may also tend to lie in other areas that are less clean than the stalls. If the stalls are too large, cows become dirty from manure dropped in the stall and more labour will be expended in cleaning the shed area. A bar placed across the top of the free stalls will prevent the cow from moving too far forward in the stall for comfortable lying down movements, and it will encourage her to take a step backwards when standing so that manure is dropped outside the stall.

Tie-stall sheds

Only in the case of purebred herds, where considerable individual attention is given to cows, can a tie-stall system be justified in tropical areas. If such a system is chosen, stalls and equipment may be purchased, in which case floor plans and elevations may be available from the equipment supplier.

The tie-and-feed-barrier construction must allow the cow free head movements while lying down as well as standing up. However, it should prevent the cow from stepping forward into the feed trough. Most types of yoke restrict the cow's movements too much. A single neck rail, set about 1 metre high and 0.2 metres over the manger may bruise the cow's neck when it pushes forward to reach the feed.

The feed barriers that best meet the requirements are shoulder supports and the comfort stall. Note the fixing rods for the cross-tie, which allows for vertical movement of the chain. Stall partitions should be used at least between every second cow to prevent cows from trampling each other's teats and to keep the cow standing straight so that the manure falls in the gutter.

Bull Pens

A bull pen should have a shaded resting area of 12–15 square metres and a large exercise area of 20–30 square metres. The walls of the pen must be strong. Eight horizontal rails made of minimum 100 mm round timber or 50 mm galvanized steel tubes to a total height of 1.5 metres, fixed to 200 mm timber posts not more than 2 metres apart, will be sufficient.

The gate must be designed so that the bull cannot lift it off its hinges, and there should be at least two exits where the herd worker can escape. A service stall where the cow can be tethered prior to, and during, service is usually provided close to the bull pen. The stall can have ramps at the sides to support the bull's front feet.

Calf Pens

Calf mortality is often high in tropical countries, but proper management and suitable housing that protects the calf from climatic stress, infections and parasites can reduce this. Individual pens for calves from birth to two to three months of age are often built with an elevated slatted floor. This floor, which is best constructed from sawn timber boards measuring 37–50 mm by 75–100 mm, leaving a 25–30 mm slit between each board, will ensure that the calf is always dry and clean.

The required minimum internal dimensions for an individual calf pen are 1 200 mm by 800 mm for a pen where the calf is kept to two weeks of age¸1 200 mm by 1 000 mm where the calf is kept to six to eight weeks of age; and 1 500 mm by 1 200 mm where the calf is kept from 6 to 14 weeks of age. Three sides of the pens should be enclosed to prevent contact with other calves and to prevent draughts.

Draughts through the slatted floor may be prevented by covering the floor with litter until the calf is at least one month of age. The front of the pen should be made so that the calf can be fed milk, concentrates and water easily from buckets or a trough fixed to the outside of the pen, and so that the calf can be moved out of the pen without lifting. The milk or milk substitute fed to the calf will not provide it with enough liquid and therefore it should be given fresh, clean water daily, or preferably have continuous access to water in a drinking nipple.

All calves, but especially those that are weaned early, should have access to good-quality forage as soon as possible to stimulate rumen development. Forage can be supplied in a rack placed above the side wall of the pen. Calf pens are recommended where the cows are kept in a semizero-grazing or zero-grazing system.

Another system that works well is the use of individual hutches. The hutch must be thoroughly cleaned and set up in a new location every time a new calf is housed in it. Plenty of litter is placed directly on the ground inside the hutch. Protection from wind, rain and sun is all the calf requires, but the key to success is to always move the hutch to clean ground.

Housing for the Small Herd

For the smallholder who wants to make the very best use of his crop land and to provide his cattle with good housing that will encourage high production, a zero-grazing system is recommended. Gum-poles may be used instead of the cedar posts and sawn rafters, but any wood in contact with, or within 50 cm of, the ground should be well treated with wood preservative. It is desirable to pave the alley but, if that is not possible, the distance between the free stalls (cubicles) and the feed trough should be doubled or tripled. A concrete pit or sloping slab in which to accumulate manure is essential. If the alley is paved, the pit can also collect urine. In fact, paving the alley not only saves space, but the value of the urine will help to pay for the paving. A water tank to collect water from the roof can be very useful unless there is an abundant supply of water nearby.

Housing for Medium to Large Herds

For the farmer with up to about 30 cows, a yard with a paved shade and feed area would be suitable. The yard and feeding area may alternatively be combined with an open-sided barn designed for deep bedding or equipped with free stalls and, where the herd consists of high-yielding cows, the milking shed may be equipped with a bucket milking machine. Some farmers with up to 30 cows may even consider using an open-sided tie-stall shed.

In general a medium- or large-scale dairy unit may include the following facilities:

1. Resting area for cows:
   (a) paved shade; or
   (b) deep bedding in an open sided barn; or
   (c) free stalls in an open-sided barn.
2. Exercise yard (paved or unpaved).
3. Paved feed area:
   (a) fence-line feed trough (shaded or unshaded); or
   (b) self-feeding from a silage clamp.
4. Milking Centre:
   (a) milking shed or parlour; and
   (b) collecting yard (part of the exercise yard); and
   (c) dairy, including milk store; and
   (d) motor room.
5. Bull pen with a service stall.
6. Calving pen(s).
7. Calf accommodation.
8. Young stock accommodation (yard with paved shade and feed area).
9. Bulk feed store (hay and silage).
10. Concentrate feed store.
11. Veterinary facilities:
    (a) diversion pen with artificial insemination stalls; and
    (b) isolation pen.

12. Waste store:
    (a) slurry storage; or
    (b) separate storage of solids and effluents.
13. Office and staff facilities.

Each of the parts of the dairy unit may be planned in many different ways to suit the production management system and the chosen method of feeding. Some requirements and work routines to consider when the layout is planned are:

1. Movement of cattle for feeding, milking and perhaps to pasture.
2. Movement of bulk feed from store to feeding area, and concentrates from store to milking shed or parlour.
3. Transfer of milk from milking shed or parlour to dairy and then off the farm. Clean and dirty activities, such as milk handling and waste disposal, should be separated as far as possible.
4. The diversion pen, with artificial insemination stalls and any bull pens, should be close to the milking centre because any symptoms of heat or illness are commonly discovered during milking, and cows are easily separated from the rest of the herd when leaving the milking area.
5. Easy and periodical cleaning of accommodation, yards, milking facilities and dairy, and transfer of the waste to storage and then to the fields.
6. Movements of herd workers. Minimum travel to move cows in or out of the milking area.
7. Provision for future expansion of the various parts of the unit.

Feeding Equipment

One advantage of loose housing of cattle is the opportunity to construct the feed trough in the fence to allow easy access for filling. The simplest type of manger consists of a low barrier with a rail fixed above. The drawback is that cattle have a tendency to throw feed forward while eating, but a wall in front.

The dimensions of the trough must be chosen to conform with the required height, reach and width of the feeding space for the animals to be fed, while providing enough volume for the amount of feed distributed at each feeding time.

Although timber construction is simple to install, concrete should be considered because of its greater durability. When timber is used, the base should be well treated with wood preservative. However, the preservative should not be used on any surface that cattle can reach to lick, as some preservative materials are toxic to animals.

When concrete is used, the grade should be at least C20, or a nominal mix of 1:2:4, because a lower grade concrete would soon deteriorate as a result of chemical attack by feedstuffs and the cow's saliva. The cows will press against the barrier before and during feeding, so the head rail must be firmly fixed to the vertical posts, which are set immovably in the ground.

A 2.5 metre-wide concrete apron along the feed trough will reduce the accumulation of mud. A narrowstep next to the trough will help to keep the trough free from manure, as animals will not back onto such a step. The bottom of the feed trough should be at a level 100 mm to 400 mm above the level at which the cow is standing with her front feet.

A slightly more elaborate feed trough separates the cattle by vertical rails or tombstone barriers to reduce competition during eating. The tombstone barrier may also reduce fodder spillage because the cow has to lift her head before withdrawing it from the trough.

A simple roof constructed over the feed trough and the area where the cows stand to eat will provide shade and encourage daytime feeding in bright weather, while protecting the feed from water damage in rainy periods.

Watering Equipment

Drinking-water for cattle must be clean. Impurities may disturb the microbiological activities in the rumen. In dairy cows, the need for water will increase with milk yield.

Water Troughs

The size of a water trough depends on whether the herd is taken for watering periodically or is given water on a continuous basis. If water is limited, the length of the trough should be such that all the cows can drink at the same time. A trough space of between 60 cm and 70 cm should be allowed for each cow. For free choice, the trough should be sized for two to three cows at a time. One trough should be provided for every 50 animals. The length may be increased if necessary. A float valve installed on the water supply pipe controls the level automatically. A minimum flow rate of 5–8 litres per minute for each cow drinking at any one time is desirable. To prevent contamination of the water trough with manure, the trough should preferably have a 300–400 mm-wide step along the front. The animals will readily step up to drink, but will not back onto the raised area. An alternative is to make the sides facing the cattle sloping. Young stock in a loose housing system require one water trough for every 50–60 animals. A 60 cm height is satisfactory. A minimum flow rate of 4–5 litres per minute for each animal drinking at any one time is desirable.

Automatic Drinkers

Automatic drinkers activated by the animals provide a hygienic means of supplying water for cows and young stock. When used in loose housing systems for cows, the bowl should be placed at a height of 100 cm and be protected by a raised area beneath it (1 metre wide and 150–200 mm in height). One bowl should be provided for every 10–15 cows.

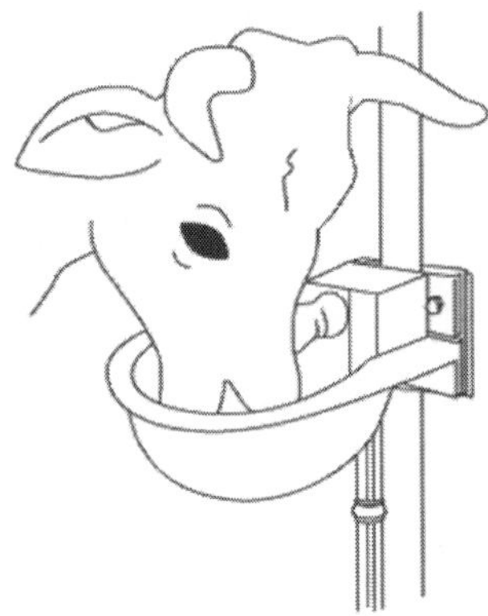

Figure 1: Automatic water drinkers

A nipple drinker without a bowl provides the most hygienic means of watering for young stock, but most nipples have a limited flow rate and can therefore not be used for calves older than six months.

Feed Handling

The types and quantities of feedstuffs to be handled vary greatly from farm to farm.

Dry Hay or Forage

If an adequate supply of green forage can be grown throughout the year, then only temporary storage and space for chopping is required. On the other hand, if a prolonged dry season makes it necessary to conserve dry forage, a storage method that will prevent spoilage is essential. A raised slatted floor with a thatched or corrugated steel roof will provide good protection for hay. If the store is filled gradually, it may help to have some poles in the top of the shed on which to spread hay for final drying before it is packed into the store. Loose hay weighs about 60–70 kg/m^3. Although requirements will vary greatly, a rough guide is 3–5 kilograms of hay or other forage per animal per day of storage.

Silage

Good-quality silage is an excellent feed for cattle. However, it is not practical for the smallholder with only a few cows because it is difficult to make small quantities of silage without excessive spoilage.

Successful silage-making starts with the right crop. The entire maize plant, including the grain, is ideal, as it has enough starch and sugar to ferment well. In contrast, many grasses and legumes do not ferment well unless a preservative, such as molasses, is added as the forage is placed into the silo.It takes a good silo to make good silage. The walls must be smooth, airtight and, for a horizontal silo, the walls should slope about 1:4 so that the silage packs tighter as it settles. The forage to be made into silage should have a moisture content of about 30–50 percent and must be chopped finely and then packed tightly into the silo. The freshly placed material must be covered and sealed with a plastic sheet. Failure at any step along the way spells disaster. The large commercial farmer, with well constructed horizontal or tower silos and the equipment to fill them, has the chance to make excellent feed. However, good management is no less important, regardless of size.

Cattle Dips

Ticks continue to be one of the most harmful livestock pests in east Africa. As vectors of animal disease, ticks have been a great hindrance to livestock development, especially in areas where breeds of cattle exotic to the environment have been introduced. At present the only effective method of control for most of these diseases is to control the vector, i.e. the ticks. Dipping or spraying with an acaricide is the most efficient way of reducing the number of ticks.

Siting a Dip

The ground where a dip is to be built, and the area around it, should be slightly sloping and as hard as possible, but not so rocky that a hole for the dip cannot be dug. Laterite (murram) soil is ideal. The ground must support the structure of the dip, be well drained and not become muddy in wet weather. It must also be resistant to erosion or gullying of cattle tracks.

Cattle must not be hot or thirsty when they are dipped, so it is important to have a water trough inside the collecting yard fence.

Waste Disposal and Pollution

All dipping tanks need to be cleaned out from time to time with disposal of the accumulated sediment. It is normal for all the waste dip-wash to be thrown into a 'waste pit' that is dug close to the dip. In addition, dipping tanks may crack with the resulting leakage of acaricide.

The dip and the waste pit must therefore be sited to ensure that there is no risk of acaricide polluting drinking-water supplies, either by overflowing or by percolating through the ground. The waste pit should be at least 50 metres from any river or stream, 100 metres from a spring or well, and considerably more than that if the subsoil is sandy or porous.

Footbaths

Footbaths are provided to wash mud off the feet of the cattle to help keep the dip clean. At least two are recommended, each 4.5 metres long and 25–30 cm deep but in muddy areas it is desirable to have more. Up to 30 metres total length may sometimes be required. The floors of footbaths should be studded with hard stones set into the concrete to provide grip, and to splay the hoofs apart to loosen any mud between them.

The footbaths should be arranged in a cascade, so that clean water added continuously at the end near the dip overflows from each bath into the one before it, with an overflow outlet to the side near the collecting pen. Floor-level outlet pipes from each bath can be opened for cleaning. If the supply of water is extremely limited, footbath water can be collected in settling tanks and reused later.

Jumping Place

A narrow, steep flight of short steps ensures that:

- animals can grip and jump centrally into the dip;

- their heads are lower than their rumps at take-off;
- they jump one at a time;
- dip-wash splashing backwards returns to the dip.

The lip of the jumping place experiences extreme wear and should be reinforced with a length of 10 cm-diameter steel pipe.

Splash walls and ceiling are provided to catch the splash and prevent the loss of any acaricide. The ceiling protects the galvanized roof from corrosion. The walls can be made of wood, but masonry is more durable.

The Dipping Tank

The dipping tank is designed to a size and shape to fit a jumping cow and allow her to climb out, while economizing as far as possible on the cost of construction and the recurrent cost of acaricide for refilling. A longer tank is needed if an operator standing on the side is to have a good chance of reimmersing the heads of the animals while they are swimming, and increased volume can slightly prolong the time until the dip must be cleaned out. In areas with cattle of the 'Ankole' type with very long horns the dip-tank needs to be much wider at the top.

Poured reinforced concrete is the best material to use in constructing a dipping tank in any type of soil. While it is expensive if only a single tank is to be built because of the cost of the formwork involved, the forms can be reused. If five tanks are built with one set of forms, the cost per tank is less than the cost of building with other materials, such as concrete blocks or bricks. A reinforced concrete dipping tank is the only type with a good chance of surviving without cracking in unstable ground. In areas prone to earthquakes, a one-piece tank is essential.

Catwalks and handrails are provided to allow a person to walk between the splash walls to rescue an animal in difficulty. In addition to providing shade, a roof over the dipping tank reduces evaporation of the dip-wash, prevents dilution of the dip-wash by rain, and in many cases, collects rainwater for storage in a tank for subsequent use in the dip.

Draining Race

The return of surplus dip-wash to the dipping tank depends on a smooth, watertight, sloping floor in the draining race. A double race reduces the length and is slightly cheaper in materials, but a very long single race is preferable where large numbers of cattle are being dipped.

Side-sloping of the standing area towards a channel or gutter increases the backflow rate. The total standing area of the draining race is the factor that limits the number of cattle that can be dipped per hour, and the size shown in the drawings should be taken as the minimum.A silt trap allows settling of some of the mud and dung from the dip-wash flowing back to the tank from the draining race. The inlet and outlet should be arranged so that there is no direct crossflow. Provision must be made to divert rainwater away from the dip.

Cattle Spray Race

A spray race site requires the same features as a dip site and these have already been described. The only difference is that the dip tank has been changed for a spray race. The race consists of an approximately 6-metre long and 1-metre wide tunnel with masonry side walls and a concrete floor. A spray-pipe system with a length of 3–3.5 metres in the tunnel with 25–30 nozzles placed in the walls, ceiling and floor, discharges dip liquid at high pressure and exposes the cattle passing through to a dense spray.

The fluid is circulated by a centrifugal pump giving a flow of 800 litres per minute at 1.4 kg/cm^2 pressure. Power for the pump can be supplied by a 6-horsepower to an 8-horsepower stationary engine, a tractor power take-off, or a 5-horsepower electric motor. The discharged fluid collected on the floor of the tunnel and draining race is led to a sump and recirculated.

In addition to being cheaper to install than a dipping tank, the spray race uses less liquid per animal and operates with a smaller quantity of wash, which can be freshly made up each day. Spraying is quicker than dipping and causes less disturbance to the animals. However, spray may not reach all parts of the body efficiently or penetrate long hair. The mechanical equipment used requires power, maintenance and spare parts, and the nozzles tend to become clogged and damaged by horns. Hand-spraying is an alternative method that can work well if carried out by an experienced person on an animal properly secured in a crush. The cost of the necessary equipment is low, but the consumption of liquid is high as it is not recirculated. The method is time-consuming and therefore only practicable for small herds where there is no communal dip tank or spray race.

Pig Housing

Pig farming is relatively unimportant in most regions of Africa, as in most tropical countries except China and South-East Asia. However, pig production is increasing in many tropical countries as processed pork finds an increasing market, and pig production yields a relatively rapid rate of return on the capital employed. Pigs are kept primarily for meat production, but the by-products, such as pigskin, bristles and manure, are also of economic importance. To some extent, pigs compete with humans for food, but they can also utilize by-products and human food waste.

Management Systems in Intensive Commercial Pig Production

There is no standard type or system of housing for pigs. Instead, accommodation and equipment are selected to suit the type of management system adopted. However, there are certain similar principles and practices in most systems. These stem from the fact that most pig units will contain pigs of different ages and classes.

Farrowing/Suckling Pens

In small- and medium-scale intensive pig production units, a combined farrowing, suckling and rearing pen is normally used. The sow is brought to this pen one week before farrowing and stays there, together with her litter, for five to eight weeks, when the piglets are weaned by removing the sow. The sow is often confined in a farrowing crate a few days before, and up to a week after, birth to reduce piglet mortality caused by overlaying or trampling.

Early weaning after a suckling period of five to six weeks, or even less, can only be recommended where management and housing is of good standard.

The piglets remain in the farrowing pen after weaning and until they are 12–14 weeks of age, or weigh 25–30 kilograms.

Group keeping of farrowing/suckling sows that have given birth within a two- to three-week interval is possible, but is unusual in intensive production. However, there are few acceptance problems, and the litters cross-suckle and mix freely. The pen should have at least 6 square metres of deep litter bedding per sow, with an additional creep area of 1 metre.

In a large-scale unit with a separate farrowing house, sometimes one of the following two alternative systems is used instead of the system described before.

The first alternative is similar to the system already described, but the piglets are moved two weeks after weaning to a weaner pen, where they may remain either until they are 12–14 weeks of age (25–30 kilograms) or until 18–20 weeks of age (45–55 kilograms). Note that the piglets should always remain in the farrowing/suckling pen for a further one to two weeks after the sow has been removed to avoid subjecting them to any new environmental or disease stress while they are being weaned. The weaning pens can contain one litter or 30–40 pigs. The pigs are often fed *ad libitum*.

In the second alternative, the sow is placed in a farrowing crate in a small pen one week prior to birth. Two weeks after farrowing, the sow and the litter are moved to a larger suckling pen. The piglets may remain in this pen until 12–14 weeks of age, or they are transferred to weaner accommodation two weeks after weaning.

Dry Sow Pens

After weaning, a sow will normally come on heat within five to seven days and thereafter at three-week intervals until successful mating. The average weaning-to-conception interval can vary between 8 days and 20 days depending on management. If the period until pregnancy has been ascertained, the sow is best kept in a pen or stall in close proximity to the boar pen.

Gestating sows are kept in yards or pens in groups of 10–12 sows that will farrow within a two- to three-week interval. They can also be kept in individual pens confined in stalls or tethered in stalls.

Weaner and Fattening Pens

The weaners, whether they come from a farrowing pen or a weaner pen, at 12–14 weeks of age will be sufficiently hardened to go to a growing/finishing pen. Finishing can be accomplished either in one stage in a growing/finishing pen from 25 kg to 90 kg– or in two stages so that the pigs are kept in a smaller growing pen until they weigh 50–60 kg and are then moved to a larger finishing pen, where they remain until they reach marketable weight.

In large-scale production, the pigs are arranged into groups of equal size and sex when moved into the growing/finishing pen. Although finishing pigs are sometimes kept in groups of 30 or more, pigs in a group of 9–12, or even less, show better growth performance in intensive systems. An alternative, where growing and finishing are carried out in the same facility, is to start about 12 pigs in the pen and later, during the finishing period, reduce the number to nine by taking out the biggest or smallest pigs from each pen.

Replacement Pens

In intensive systems a sow will, on average, produce three to six litters before it is culled owing to infertility, low productivity or age. Young breeding stock should be separated from the rest of the litter at about three months of age, because they should be less intensively fed than the fattening pigs.

Gilts are first mated when they are seven to nine months of age, or weigh 105–120 kg. After mating, they can either be kept in the same pen up to one week before farrowing, or kept in the gestating sow accommodation, but in a separate group.

Boars in the tropics are usually quiet if run with other boars or with pregnant sows, but may develop vicious habits if shut up alone.

Determining the Number of Pens and Stalls Required in A Pig Unit

One objective of planning a pig unit is to balance the accommodation between the various ages and numbers of pigs. Ideally, each pen should be fully occupied at all times, allowing only for a cleaning and sanitation period of about seven days between successive groups. In the following example, we determine the number of different pens required in a 14-sow herd where eight-week weaning is practiced.

Space Requirement

In intensive pig production systems, all pigs should be raised on concrete floors to ensure a clean and sanitary environment. In semi-intensive systems, a concrete floor is used only in the pens for finishing pigs and perhaps in the farrowing pens, whereas an earth floor or deep-litter bedding is used in other pens and yards. Litter may or may not be used on a concrete floor, but its use is desirable, particularly in farrowing pens.

The cost of a concrete floor is relatively high, resulting in a tendency to reduce the floor area allowed per animal. However, excessively high stocking densities could retard performance, increase mortality, health and fertility problems, and result in a high frequency of abnormal behaviour, endangering the welfare of the animals. An increase in the stocking density must be accompanied by an increased standard of management and more efficient ventilation and cooling.

In particular, to aid cooling, finishing pigs kept in a warm tropical climate should be allowed more space in their resting area than is normally recommended for pigs in temperate climates.

The dimensions of a pen for fattening pigs are largely given by the minimum trough length required per pig at the end of their stay in the pen. However, the width of a pen with low stocking density can be larger than the required trough length. This will reduce the depth to 2.0–2.4 metres, and run the risk of having the pigs create manure within the pen.

Furthermore, this increases flexibility in the use of the pen and the extra trough space allows additional animals to be accommodated temporarily or when the level of management improves.

Sometimes finishing pens are deliberately overstocked. The reason for this is that all pigs in the pen will not reach marketable weight at the same time and the space left by the pigs sent for slaughter can be utilized by the remainder. Such overstocking should be practiced only in very well managed finishing units.

General Requirements for Pig Housing

A good location for a pig unit meets the following requirements: easy access to a good all-weather road; well-drained ground; and sufficient distance from residential areas to avoid creating a nuisance from odour and flies.

An east-west orientation is usually preferable to minimize exposure to the sun. Breezes across the building in summer weather are highly desirable. A prevailing wind during hot weather can sometimes justify a slight deviation from the east-west orientation. Ground cover, such as bushes and grass, reduces the reflected heat considerably, and the building should be located where it can benefit most from surrounding vegetation. A fairly light, well-drained soil is preferable, and usually the highest part of the site should be selected for construction.

Pig houses should be simple, open-sided structures because maximum ventilation is needed. A building for open confinement is therefore essentially a roof carried on poles. The roof supporting poles are placed in the corners of the pens where they will cause least inconvenience. A free-span trussed roof design would be an advantage but is more expensive.

In some circumstances it may be preferable to have solid gable ends and one enclosed side to give protection from wind or low temperatures, at least for part of the year. If such walls are needed, they can often be temporary and be removed during hot weather to allow maximum ventilation. Permanent walls must be provided with large openings to ensure sufficient air circulation in hot weather. If there is not sufficient wind to create a draught in hot weather, ceiling fans can improve the environment considerably.

The main purpose of the building is to provide shade, and therefore the radiant heat from the sun should be reduced as much as possible. In climates where a clear sky predominates, a high building of three metres or more under the eaves gives more efficient shade than a low building. A wide roof overhang is necessary to ensure shade and to protect the animals from rain.

A shaded ventilation opening along the ridge will provide an escape for the hot air accumulating under the roof. If made from a hard material, the roof can be painted white to reduce the intensity of solar radiation. Some materials, such as aluminium, reflect heat well provided that they are not too oxidized.

A layer of thatch (5 cm), attached by wire netting beneath a galvanized steel roof, will improve the microclimate in the pens. A roof of thatch is excellent in hot climates, particularly in non-confined systems, but cannot always be used because of the fire hazard and because it attracts birds and rodents. A pig house with two rows of pens and a central feeding alley would require a ridge height of 5–6 metres if covered with thatch.

The pen partitions and the 1-metre wall surrounding the building, which serves to reduce heat reflected from the surrounding ground, can be made of concrete blocks or burnt clay bricks for durability, or perhaps soil-cement blocks, plastered for ease of cleaning. Regular whitewashing may improve the sanitary conditions in the pens.

Doors have to be tight-fitting and any further openings in the lower part of the wall surrounding the building should be avoided in order to exclude rats. Apart from stealing feed and spreading disease, large rats can kill piglets.

For all types of confinement housing, a properly constructed, easily cleaned concrete floor is required. A 80–100 mm layer of concrete on a consolidated gravel base is sufficient to provide a good floor. A stiff mix of 1:2:4 or 1:3:5 concrete, finished with a wood float, will give a durable nonslip floor. The pen floors should slope 2–3 percent towards the manure alley, and the floor in the manure alley should slope 3–5 percent towards the drains.

Housing for a Small-scale Pig Unit

For units with 2–15 sows, specialized buildings for the various stages of production may not be practical or desirable. For the smallest units of two to six sows, a kind of universal pen can be erected about 2.7 metres wide and 2.8–3.0 metres deep (including feed trough), which can be used for:

1. one sow and her litter; or
2. one litter of weaned piglets; or
3. up to four gestating sows; or
4. growing/finishing pigs up to 90 kg live weight or 1 boar.

This type of pen provides a high degree of flexibility but usually does not allow such efficient use of the building space as the more specialized pens.

When used for farrowing, the pen should be adapted with guard rails 25 cm above floor level and 25 cm from the wall to protect the piglets from being crushed, However, confinement farrowing is one of the most efficient ways of reducing piglet losses. An arrangement with fixed or removable rails, which divide the pen, will offer some degree of confinement.

In some climates it may be desirable to give sows with litter access to exercise yards. However, for the relatively short suckling period (six to eight weeks), it is usually considered best to keep the sows confined in pens with their litters.

A creep for the piglets is arranged in one corner of the pen. It is recommended to construct a temporary ceiling (e.g. wire netting covered with straw) 50–60 cm above the floor in the creep area to prevent draughts and to ensure warmer temperatures for the piglets during their first weeks of life. Where electricity is available, heating with an infrared lamp may be used instead. Piglets are fed in the creep area out of reach of the sow.

The semicovered manure alleys are arranged along the outside walls, separated from the resting area of the pen. This arrangement allows rainwater to help flush away the waste to the drain channel and on to the manure store, which needs to have extra capacity for this water. However, in the four-sow unit the furrowing pens have fully covered manure alleys for increased protection of the piglets.

The roof may be equipped with gutters so that rainwater can be drained away separately or collected for use as drinking-water for the pigs.

A single tubular steel or round timber rail 20 cm above the outside rear wall (1 metre high) is desirable to increase security without interfering with ventilation.

Housing for the Large-scale Pig Unit

In large scale units, special provisions must be made for efficient health control. This means: not too many animals in one building; animals of approximately the same age housed together; using an 'all-in, all-out' system with thorough cleaning and disinfection of every house between each batch of pigs; placing the buildings 15–20 metres apart and surrounding the entire site with a secure fence.

Specialized pens located in separate houses assigned to the various stages in the production cycle are normally feasible in units of 20–30 sows. Each type of pen can be designed with dimensions for the most efficient use of the building space, as it is not necessary for them to fit in a layout with other types.

Farrowing House

Farrowing pen offers a relatively high degree of confinement, as the sow is restrained in a farrowing crate during farrowing. Between 5 and 10 days after farrowing, the crate is removed or opened to free the sow. Although a slightly askew arrangement of the farrowing crate will allow for a longer trough for the piglets in the front of the pen, it is more complicated to construct. A reduction in space requirements can be achieved by putting the sow in a farrowing pen, consisting merely of a farrowing crate with 0.5 metre- and 1 metre-wide creep areas on either side, one week prior to farrowing. Two weeks after farrowing, the sow and piglets must be transferred to a suckling pen equipped, but with the dimensions 2.3 metres wide by 2.35 metres deep and with a 1.4-metre wide manure alley.

Housing for Growing/finishing Pigs

Growing/finishing pens 2.8 metres wide by 1.9– 2.2 metres deep and with a 1.2-metre manure alley can accommodate the following number of pigs, according to their weight:

up to 40 kg – 12 pigs 40–90 kg – 9 pigs over 90 kg – 7 pigs

Where it is very hot, it is preferable to reduce the number of pigs per pen below the numbers given here. The manure alley must be well drained, preferably by a covered drain, but an open drain will also serve, provided that it is outside the pen to prevent urine from flowing from one pen to another. Bedding in the pens is preferable for the animals' comfort and to reduce stress, as the bedding will provide them with something to do. Controlled feeding is important to ensure the best possible feed conversion.

Housing for Gestating Sows

Gestating sows are usually the last group in a pig herd to be considered for confinement housing. However, there are obvious advantages in doing so, which could greatly influence production efficiency when sows are confined and controlled during gestation.

As their litters are weaned, sows can be returned to the gestating-sow structure and placed in one of the pens arranged on either side of the boar pens for easy management of sows in heat. After mating and the three-week control period, the sows should be regrouped according to the actual farrowing dates.

Behind the feeding stalls there is a manure alley with gates across, which can close the opening of the resting area in order to confine the sow while cleaning out the manure alley. The width of the manure alley can be increased from 1.5 metres to 2.5 metres if desired, so that cleaning can be carried out by a tractor-mounted scraper.

Where exercise yards are considered feasible, they can be arranged behind the building in both types of pen.

5 litres per pig per day using a wallow. However, a spray system can be operated intermittently by a timer that can limit use to about 2 litres per pig per day. The spray should be directed onto the pigs and not into the air.

Special Arrangements for Warm Climates

Many of the principles discussed above apply equally to hot and temperate climates and are basic requirements for the housing of pigs. While the open type of confinement system has its limitations, when applied in many warm areas it leads to a major improvement in production.

Complete control of the environment in animal houses is generally far too expensive to be feasible, in particular for non-confined systems. However, provisions for shade, proper roof colour and material, and controlled air movements, which have already been discussed, can be both practical and economic.

A spray or a wallow can considerably reduce heat stress in pigs. A wallow can be anything from a water-filled hole in the earth to a concrete trough. While wallows are effective and need not be expensive, they tend to become unsanitary if not regularly cleaned.

From a hygienic point of view, sprinklers that spray water onto the pigs are preferable, but water consumption can be up to four times greater than with a wallow. Water consumption is about 20 litres per pig per day for 10 hours of continuous spraying, compared with

The spray system can be effectively used with all categories of pig, except very young piglets. A sprinkler in the manure alley of the farrowing pen, operated from the time the litter is about two weeks old, may help the sow to maintain her feed intake. Hosing pigs once or twice a day is a great deal less effective than a spraying system.

Feed Troughs and Feed Storage

Efficient pig production requires a reliable supply of water and feed for a balanced diet. A large range of feedstuffs, including by-products and crop surpluses, may be used, provided they are incorporated into a balanced diet. Feed requirements change as pigs grow and depend on the stage of production in sows. A wide variety of feeding equipment is available for pig operations. The easiest to clean and sanitize are made from concrete, metal or glazed burnt clay.Concrete troughs are commonly used and can be prefabricated using a metal mould. The trough is often placed in the front wall of the pen.

Although such an arrangement involves more difficult construction than having the trough inside the pen it is usually preferred for easier feeding and also preventing the pigs from stepping into the trough. The wall above the trough can be made either solid or open and can be either vertical or sloping inwards to the pen.

An open front improves ventilation in the pen but it is more expensive than a close-boarded wooden front as galvanized steel pipes have to be used for durability. In particular a sow confined in a stall of a farrowing crate will feel more comfortable if it is able to see in front of her. A sloping front will more effectively discourage pigs from stepping into the trough but it is more complicated and expensive to construct.

Watering Equipment

It is preferable to mix feed meal with 1.5–2.1 litres of water per kilogram of feed. The rest of the water can be given in the trough between feedings or in special drinkers. Clean water must be available to the pigs at all times, including the piglets in a farrowing pen. Automatic drinkers are the most hygienic and can be used where piped water is available. There are two types: one is placed above the feed trough and sprays into the trough when pushed by the pigs; the other type is operated by the pigs biting around it. This latter type is often placed in the manure alley or in the pen close to the manure alley to prevent the pigs from making the resting area wet.

Manure Handling

The pig pens must be cleaned once or twice a day. Provided sufficient bedding is used and the urine is drained away separately to a urine storage tank, the solids may have a consistency that allows stacking on a concrete slab.

Poultry Housing

Poultry (which includes chickens, turkeys, ducks and geese) offers one of the best sources of animal protein, in the form of both meat and eggs, at a cost most people can afford. Chickens are the most widely raised and are suitable even for the smallholder who keeps a few birds that largely forage for themselves and require minimum protection at night.

At the other extreme, commercial farms may have highly mechanized systems housing thousands of birds supplying eggs and meat to the city market. In between are farm operations in a wide range of sizes, with varying types of housing and management systems proportionate to the level of investment and the supply of skilled labour available.

No single system of housing is best for all circumstances, or even for one situation. Some compromise will invariably be required. Here we discuss the needs of chickens and other classes of poultry and a number of housing systems, along with the principal advantages and disadvantages of each.

General Housing Requirements for Chickens

Proper planning of housing facilities for a flock of laying hens requires knowledge of management and environmental needs during the various stages of the chicken's life.

The laying period may be up to 16 months but, in flocks kept for commercial egg production, the hens are normally culled after a laying period of 11–12 months, or when production has dropped to a point where the number of eggs collected per day is about 65 percent of the number of hens in the flock.

The hens may come into production again after a moulting period of a couple of months, but the production then is not as high and the egg quality is generally not quite as good as in the first laying period. Where prices of poultry meat are reasonable, it is usually more economic to cull all the hens after one year of egg production.

Site Selection

The best site is one that is well drained, elevated but fairly level, and has an adequate supply of drinking-water nearby.

Regardless of the type or size of the housing system, the site for construction should be selected to provide adequate ventilation, but be protected from strong winds. An area under cultivation, producing low-growing crops, will be slightly cooler than an area of bare ground. High trees can provide shade while at the same time actually increasing ground-level breezes. Bushes planted at one windward corner and also at the diagonally opposite corner will induce air currents within the building to reduce the effect of the heat from direct solar radiation.

As all buildings used for poultry housing tend to produce odours, they should be located well downwind of nearby dwellings. If there are several poultry buildings in a group, it is desirable to have them separated by 10–15 metres in order to minimize the possibility of spreading disease.

Brooding buildings should be isolated from other poultry buildings by 30 metres or more, and be self-contained in terms of feed supplies and storage of equipment. If the same person cares for both layers and growing birds, a disinfectant foot bath at the entrance to the brooding area is an added precaution. All buildings should be constructed on well-drained sites where drives and paths between buildings will not become muddy, even during the rainy season.

Environmental Requirements

The effects of temperature and humidity on the birds make it apparent that in most areas of east and southeast Africa. the principal environmental concern is to keep the flock as cool as possible. Shade, good ventilation with natural breezes, freedom from roof radiation and the indirect radiation from bare ground are all important. Only in a few high altitude areas does protection from wind and low temperatures become a significant consideration.

Humidity seems important in only two respects. Very low humidity causes objectionably dusty conditions and high humidity, combined with temperatures above 27 °C, seems to interfere with the physiological cooling mechanism and increases the possibility of death. Day-old chicks require a temperature of 33–35 °C. This temperature is maintained for a week and is then gradually lowered to the ambient temperature by the end of five weeks.

In addition to providing a good environment, the housing should offer protection from predators and theft, as well as keeping out rodents and birds. Not only do they carry diseases, they can also consume enough feed to make a significant economic difference.

The effect of light on egg production has been discussed earlier. Additional hours of light can be achieved by installing one 40-watt electric light bulb per 15 square metres of floor space in a position about 2.2 metres above floor level. More important than the hours of light, however, is maintenance of the lighting schedule, because any sudden change in the length of the photo-period is likely to result in a significant drop in production.

Fourteen hours of light throughout the laying period is optimum. A schedule with gradually decreasing hours of light may be used in windowless houses for maturing pullets. This postpones laying but results in larger eggs being produced from the start of laying.

However, in warm climates near the equator, houses are open for natural ventilation and the length of the day is close to 12 hours throughout the year. The result is that pullets start to lay at 14–18 weeks of age and egg size, which is small at first, gradually increases during the first three months. Broiler houses are often lighted 24 hours a day to encourage maximum feed consumption and rate of gain.

Proper design and management of the poultry house can effectively contribute to disease prevention in the flock. In general, it is best if the litter is dry but not too dusty. If no litter is used, the floor and wall surfaces should be designed so that they can be cleaned easily between flocks and stay reasonably clean during use.

Construction Details

In most hot climates there will be many more days when a cooling breeze is needed rather than protection from a chilling wind. A wall construction consisting of a solid base, which protects against indirect radiation from the ground, and an open space covered with mesh above it, is therefore preferred for all four walls in most types of chicken house.

A hessian or reed curtain that can be dropped on the windward side will offer extra protection and, if installed on the east and west orientation, it may also protect from direct sunshine. An arrangement where the top end of the hessian is fixed to the wall plate and the bottom end is attached to a gum-pole, around which it can be rolled when not in use, provides for smooth operation. In high altitude areas, offcuts may be used on the gable ends, but 15–20 mm spaces should be left between them to improve ventilation. The width of the building should not exceed 9 metres for efficient cross-ventilation.

The lower wall design, up to 1 metre of solid wall, can be made of any available masonry units. Bag-washing will give a smooth, easily cleaned finish, but adobe blocks will require the extra protection of plastering to prevent the birds from destroying the wall by pecking.

The upper wall design to the total height of the wall, including the solid base, should be about 2 metres. gum-poles treated with wood preservative and set 500 mm deep in concrete provide a practical means for supporting the roof and upper wall structure. Eighteen-millimetre wire mesh is small enough to keep out rodents and birds. A tight-fitting door is essential.

The floor in a poultry house may consist of gravel or well-drained soil, but concrete is desirable because it is easy to clean, durable and considerably more rat proof. A concrete floor should be 80–100 mm thick and be made of a stiff 1:2:4 or 1:3:5 mix, laid on a firm base at least 150 mm above ground level, and given a smooth finish with a steel trowel.

Roof structures with a free span are desirable to avoid any inconvenience from roof-supporting poles inside the building. Corrugated steel sheets are the first choice for roofing material because they are much easier to keep clean than thatch. Insulation under the metal roofing will improve the environment in the house. However, a thatched roof may result in even better conditions and can be used on narrow buildings.

The roof overhang should be 500 mm or more in order to give adequate protection from sun and rain. A ventilation opening along the ridge is usually supplied in layer houses, but not in brooding houses.

Semi-intensive Systems

Semi-intensive systems are commonly used by small-scale producers and are characterized by having one or more pens in which the birds can forage on natural vegetation and insects to supplement the feed supplied. It is desirable to provide at least two runs for alternating use to avoid a build-up of disease and parasites. Each run should allow at least 10–15 square metres per hen and be fenced, but a free-range run allowing 40–80 square metres per hen will be required where the hens are expected to obtain a substantial part of their diet from foraging.

A small, simple house, which allows 0.3–0.4 square metres per bird and has a thatched roof, a littered earth 16–18 birds and can normally be handled by one person. For larger flocks, several such units will be used.

Portable units are generally more expensive than permanent houses and may decay quickly because of contact with the ground. The hens have reasonable protection against birds of prey and inclement weather, as well as parasites if the unit is not returned to the same area within 30 days.

In areas where grassland is limited, a yard deeply littered with straw and allowing only 0.4–0.7 metres per bird will provide an outdoor exercise area. This system is similar to the deep-litter system, but requires more space, a considerable amount of litter for the yard, and the fresh green food has to be carried to the birds.

Deep-litter System

Deep-litter houses, confine the birds in a building that offers good protection with a reasonable investment. If well designed, with low masonry walls set on a concrete floor and wire mesh completing the upper part of the walls, the building will keep out rats and birds.

The principal advantages of this system are easy access for feeding, watering and egg gathering, good protection and reasonable investment. The principal disadvantage is the need for high quality litter. If this is produced by the owner, it is of little significance but, if it must be purchased, it becomes an economic factor. In either case, the litter and manure must be removed periodically.

The deep-litter house can be designed up to 9 metres in width and any length that is needed. A satisfactory density is approximately 4–5 birds per square metre of floor area.

Slatted or Wire-floor System

Feeding, watering and egg-gathering are all efficiently handled from the outside. Either a double-pitch thatch roof or a single-pitch corrugated steel roof may be installed, with the eaves about 1.5 metres above the floor. If the latter is used, some insulation under the roof is desirable.

The feed troughs should be equipped with hinged covers, and rat guards should be installed at the top of each pier. The width of this type of building should be limited to about 2 metres to allow easy removal of manure and adequate wall space for feed and nests. The building should be oriented east and west and may be of any length. However, if it is more than 5 metres long, nests will need to be put on the sides, and all remaining wall space on either side used for feeders in order to allow the required 100 mm/bird.

If using a slatted floor sufficiently strong for a person to walk on, then a wider building is feasible, as feeders can be placed completely inside where the chickens have access to both sides of the trough. The floor is sectioned for easy removal during cleaning out of manure.

This type of house is said to be cooler than other types, but the building cost is high and management is more complicated.

Combination of Slatted Floor and Deep Litter

A combination deep-litter/slatted-floor house offers some advantages over a simple deep-litter house, but with some increase in investment.

Approximately half of the floor area is covered with small gum-pole slats or with wire mesh. This area is raised above the concrete floor by 0.5 metres or more to make it possible to clean under the slatted portion from the outside. Waterers and feeders are placed on the slatted area. This type of house is limited in width to 3–4 metres to enable feeders and waterers to be handled from the litter area, and manure beneath the slatted area can be easily removed from the outside without moving the slats or disturbing the birds. Although this system entails added expense for materials and labour to install the slats, the bird density can be increased to 5–7 per square metre, so there is little difference in the cost per bird. This system saves on litter, increases litter life, reduces contact between birds and manure, and allows manure removal without disturbing the hens. Ventilation is improved by the slatted floor. Perhaps the biggest disadvantage is the limited width for convenient operation and the need for some litter.

In medium- to large-scale houses of this type, the slatted floor must be made removable in sections, and at least part of it made strong enough to walk on. However, this results in increased building cost and more complicated management.

This is generally considered the maximum for this type of house and allows for a stocking density of up to eight birds per square metre. Automatic tube feeders are placed on the slatted floor. One such feeder, with a bottom diameter of 0.6 metres can serve 60–75 birds, depending on the size of the breed.

The water troughs are suspended from the ceiling. The nest boxes are doubled by arranging them back to back, and have one end resting on the slatted floor and the other suspended from the ceiling. Egg-collection can be facilitated by the use of a trolley, which is supported on a rail just below the ceiling. A tractor shovel can be used for cleaning out between batches if all furnishings and part of the end walls are made removable.

Cage or Battery Systems

Cage management of layers in very large, well-insulated, windowless buildings has become standard practice in much of Europe and the colder parts of the United States. With complete mechanization of feed, water, egg-collection, manure removal and environmental control, two to three people can care for thousands of birds.

It should be noted that a very large investment is made in order to obtain labour efficiency and ideal environmental conditions. East and southeast Africa currently have relatively low labour costs and a mild climate, which could make a mechanized cage system in an insulated building unnecessary and impractical.

Nevertheless, there are much simpler cage systems that may work very well for commercial growers in this region. These consist of rows of stair-step cages in long, narrow shelters. The thatch roof, or insulated metal roof shelter, can be completely open on the sides, with perhaps some hessian curtains in areas where cold winds are experienced. The buildings should be oriented east to west, and be designed to provide shade for the cages near the ends.

A 3.4-metre width can accommodate four cages without overlap and an alley of about 0.9 metres. While a concrete floor makes cleaning easy, smooth, hard soil is less expensive and quite satisfactory. A little loose sand or other litter spread on the soil before the manure collects will make manure removal easier. The building posts should be treated with wood preservative and be sturdy enough to support the cages. Rat guards should be installed on the posts at a height of 0.8–1 metre.

A cast-concrete central alley raised 20 cm is easy to clean and keeps manure from encroaching on the work area. Feeding and egg-collecting are easily done by hand, while watering may be either by hand or with an automatic system. It is important for the watering trough to be carefully adjusted so that all birds receive water. The simplest method of supplying water automatically, or by hand at one end, is to slope the entire building and row of cages by 10 mm per 3 metres of length.

The trough can then be attached parallel to the cages. Water must run the total length of the trough and it is inevitable that some will be wasted. Consequently a good water supply is essential.

Even though feed is distributed by hand, feed stores should be built convenient to each building to minimize carrying. Eggs can be collected directly on 'flats' stacked on a cart that is pushed down the alley. The cart can be made self-guiding by means of side rollers that follow the edge of the feed troughs or the raised central alley. Cages that are equipped with pans to catch the manure are not recommended because they restrict ventilation. Previously used cages should be considered only if they are of a suitable design, and have been carefully inspected for condition prior to purchasing.

Planning for Continuous Production

Producers who can supply their market with either eggs or meat on a regular and steady basis will undoubtedly find their produce in demand at the best market prices. Planning the poultry housing system has much to do with steady production. It is assumed that the brooder house is large enough for brooding only, and that pullets will be transferred to a laying house for growing to laying age. New chicks are started every 13 weeks, brooded for 7–8 weeks, and then transferred to the laying house. After approximately 11 weeks they will start a laying period of 52 weeks, after which they are sold and the house is cleaned and rested for two weeks before the cycle is renewed.

Five laying houses are required. At any one time, four will have layers in full production and the fifth will be either housing growing pullets or empty for cleaning. Each house is on a 65-week cycle: 11 weeks for growers; 52 weeks for layers; 2 weeks for cleaning. The brooder house is on a 13-week cycle: 7–8 weeks for brooding; 5–6 weeks for cleaning and resting.

Housing for Breeders

Breeders must be housed in one of the floor systems because cocks need to run with the hens. One cock per 5–10 hens is sufficient. Special emphasis is placed on disease control, so often a partially or completely slatted floor design is preferred. Few commercial producers breed their own replacements, but instead buy day-old chicks from a commercial hatchery.

However, most chicks of indigenous breeds are produced by natural incubation in small-scale farms. Although a hen sitting on some 8–10 eggs needs little feed and even less attention, breeding results may be improved by a cool, clean nest at ground level that is enclosed to protect the hen, and later the chicks, from insect pests, vermin and predators, and by a supply of feed and clean water.

Naturally hatched chicks are reared and protected by the broody hen and can be left undisturbed, provided that their yard is protected from predators, is of a good sanitary standard and has a supply of feed and water. Artificially incubated chicks must be started under gas-fired or oil-fired brooders to compensate for the absence of a natural mother, and to keep them warm without crowding together. If electricity is available, a 250-watt infrared lamp is a more reliable and comfortable solution, but it is also more expensive.

Housing for Pullets and Broilers

In the past, poultry meat has been derived chiefly from culled layers. This is still the main source of poultry meat in most developing countries, although there is an increasing shift towards rearing chickens specifically for meat. Broilers, the common term for meat birds, are fast-growing strains that reach market weight of 1.6 kg in 8–12 weeks. The commercial production of poultry meat is now based primarily on broilers.

In a semi-intensive system, the growing pullets may obtain part of their food by scavenging for forage, seed, etc. A fenced yard allowing 5–8 square metres per bird is preferable to open land. At least part of the yard should have shade cover, and a simple building in which the birds can be enclosed at night will be required. The building should allow 0.2 square metres per bird, have good ventilation and perches for roosting, and offer protection against predators and inclement weather. The birds should be moved at regular intervals to a different yard in order to avoid a build-up of worm infestation.

There is little difference between the system for rearing chicks to become pullet replacements for the laying flock and the system for rearing broilers for market. The same environment and housing are suitable, so they will be considered together.

Brooding and rearing are floor-managed operations. It is common practice to keep broilers or pullets in the same house from the time they are one day old: first on newspapers or thinly spread litter, and later on deep litter. When broilers are marketed at 8–10 weeks of age, or when pullets are transferred to the laying house at 16–18 weeks of age, the litter is removed so that the house can be thoroughly cleaned and disinfected. The house should therefore be designed and built to allow for easy and efficient cleaning. Pullets and broilers are not grown together owing to the difference in the length of their growing periods and differing schedules for artificial lighting.

Chicks are started in a brooder, which may be of the type discussed in the previous section, and remain there for 6–8 weeks. During this time it is desirable to conserve heat and to prevent draughts and, in this respect, the building design can be an important factor. A method that is widely used in the United States, called 'end-room' brooding, works well and seems adaptable to warm climates as well.

By taking advantage of the fact that chicks up to four weeks old require only one-third to half as much floor space as they will need later on, one end and enough of the adjacent sides are closed in tightly to provide 0.05 square metres of floor space per chick to be brooded. Offcuts, with low thermal capacity, are ideal for enclosing the wall. A hessian curtain forms the fourth (interior) wall to complete the temporary enclosure. The baby chicks can then be confined in the space around the brooder in the enclosed end of the house.

The remaining walls are covered with 18–25 mm wire mesh. At the end of the brooding period, the brooder is raised to the ceiling for storage, the hessian curtain is lifted and the chicks are allowed into the rest of the house, which should provide from 0.08 square metres/ bird for broilers, to 0.17 square metres/bird for pullets.

Depending on the maximum temperatures expected, it may be necessary to provide some ventilator openings in the enclosed walls. An adjustable gable-end ventilator is particularly desirable, as the roof will not have a ridge vent for brooding operations. If cool, breezy weather is expected, one or more of the screened sides may be equipped with hessian curtains.

Equipment and Stores

In addition to what has already been described, the chicken house will require equipment such as waterers, feeders and a feed store, and perhaps perches for roosting. Houses for floor-managed layers or breeders will require nest boxes. A store for eggs may be required in any laying house. There should be sufficient feeders and waterers for easy access (particularly important for young chicks), long enough for each bird to have its place, and with sufficient holding capacity.

Feeders

Either trough or tube feeders are used for day-old chicks, growing birds and layers, but their size must be selected to suit the birds to be fed. The number of feeders should be such that the distance to the nearest feeder does not exceed 2 metres from any point in the house. A trough should be not too wide, easy to clean and designed to prevent the hens from leaving their droppings in it.

Drinkers

An ordinary 10-litre or 15-litre bucket serves very well as a drinker for layers. If it is sunk into the floor or ground so that only about 10 centimetres shows above ground, it may be used for chicks as well. Another arrangement for chicks consists of a shallow bowl supplied with water from an upside-down bottle. Just like feeders, they are used on the floor for small chicks and on stands for older birds. The number of drinkers should be such that all chickens have access to one within a distance of 3 metres.

Automatic drinking nipples may be used for layers in cages. There should be at least one nipple for every two hens. It is desirable for every hen to have access to two nipples, as clogging of a nipple is not always easily detected.

Nest Boxes

Layers and breeders, except those managed in cage systems, should have access to nest boxes in which they can lay their eggs. The nests can be used by one or more birds at a time. Single nests commonly have the dimensions 250–300 mm wide, 300–380 mm deep and 300–350 mm high. They have a 100 mm litter-retaining board across the bottom of the opening and a perch 150–200 mm in front of the entrance.

Communal nests should have a space allowance of at least 0.09 square metres per bird. The top of the nest should be steeply sloped to prevent birds from roosting there. One nest should be supplied for every five birds in the flock.

Perches

Chickens have a natural instinct to roost in trees at night. To provide for this, perches are commonly installed in chicken houses from six to eight weeks of age or more, in particular in semi-intensive systems. Perches for young birds should have a diameter of about 35 mm and provide 0.1–0.15 square metres of space per bird, while adult birds need about 50 mm diameter and 0.2–0.3 square metres of space. The perches should be fixed to solid stands 0.6–1 metres above the floor and 0.35–0.4 metres apart, preferably placed lengthwise at the centre of the house. A deck about 200 mm underneath to collect manure is desirable.

Feed Stores

The feed stores required for a small flock are very similar to those for food grains. For commercial flocks, the type of store depends on how the feed is handled. If it is purchased in bag lots, then a masonry building with an iron roof that is secure against rodents and birds is most suitable. If feed is delivered in bulk, then one or more overhead bins, from which the feed is removed by gravity, will be convenient and safe.

The size of the store required depends entirely on the frequency and size of deliveries, but can be estimated as 0.0035 m^2 floor area per bird in the flock where feed is purchased in bags.

Sheep and Goat Housing

Sheep and goats are important sources of milk and meat. Both readily adapt to a wide range of climates and available feed supplies. They also have similar housing requirements and will therefore be treated together.

Management Systems

Depending primarily on the availability and use of land, three systems of production are practiced:

1. Subsistence, in which a few animals are tethered during the day and put into a protective shelter at night.
2. Extensive, in which the flock/herd grazes over large areas of marginal land unsuitable for agriculture. The flock is usually shut into a yard at night. Both these systems are practiced extensively in East Africa.
3. Intensive, in which the animals are confined to yards and shelters, and feed is brought to the flock. This system offers the greatest protection for the flock from both predators and parasites. Although it may make the best use of limited land resources, this system also increases labour and the capital investment required for facilities.

Housing

Housing in tropical and semitropical regions should be kept to a minimum, except for the more intensive systems of production. In the arid tropics, no protection other than natural shade is required. In humid climates, a simple thatched shelter will provide shade and protection from excessive rain. As sheep and goats do not tolerate mud well, yards and shelters should be built only on well-drained ground.

Unless predators are a serious problem, gum-poles can be substituted for the brick walls. If thatching is difficult to obtain, a lower-pitch roof of galvanized steel is feasible, but some insulation under the roof is desirable.

Where housing facilities are provided, in addition to water, feed troughs and permanent partitions, it will be necessary to provide temporary panels to help divide and handle the flock when necessary, in order to carry out such operations as disease treatment, docking, shearing, milking and lambing.

In temperate climates and at high altitudes, a more substantial structure may be needed. An open-front building facing north provides wind protection and a maximum of sunshine. A rammed-earth floor with a slope of 1:50 towards the open front is recommended. A concrete apron sloped 1:25 and extending from 1.2 metres inside to 2.4 metres outside will help maintain clean conditions in the barn.

Milking can be facilitated by providing a platform along the feeding fence where the animals can stand while being milked from behind. Such a platform should be 0.8 metres deep and elevated 0.35–0.5 metres above the floor where the milker stands.

Parasite Control

A dipping tank and crush are essential in the layout for a large flock, or for a community facility for the use of many smallholders.

References

Agricultural Information Centre. *Livestock development technical handbook.* Nairobi, Kenya.

Devendra, C. & Fuller, M.F. (1979). *Pig production in the Tropics.* Oxford University Press.

Francis, P. (1984). *Poultry production in the Tropics.* Intermediate Tropical Agriculture Series. London, Longman Group Ltd

Hafez, E.S.E. (1975). *The behavior of domestic animals.* 3rd edition. Baltimore, The Williams and Wilkins Co

Midwest Plan Service. (1984). *Small farms – livestock buildings and equipment.* Ames, Iowa.

Williamson, G. & Payne, W.J.A. (1978). *An introduction to animal husbandry in the Tropics.* 3rd edition. Tropical Agriculture Series. London, Longman Group Ltd.

3

Animal Feed and Nutrition

Proper feeding and nutrition is the foundation for a productive livestock sytems. Without good nutrition, animal cannot express their full genetic potential nor will they be reproductively efficient. Often low reproductive rates, poor growth, and increased illness are a result of a nutritional imbalance or deficiency rather than a disease or genetics. In addition, pasture and feed represent the single largest cost associated with the cowherd.

Livestock producers need to understand basic digestive physiology, types of nutrients, and requirements of the cow in order to be competent on-farm nutritionists. By understanding feeds and ration balancing, producers can meet the nutritional needs of their animals in a more cost efficient manner. In addition, a fundamental understanding of feeds and rations will assist producers in evaluating new products, alternative feeds, and supplements.

Cattle Have a Unique Digestive System

Cattle belong to a class of animals known as ruminants. Ruminants are cloven hooved animals that have four compartments to their stomach and chew their cud. In addition, ruminants have an unusual configuration of teeth. Their small and large intestine are designed to handle large volumes of material. Cattle evolved to exist on large amounts of fiber. They do not do well on all grain or high fat diets.

The mouths of cattle are very different from most non-ruminant animals. Cattle have 32 teeth. They have 6 incisors and 2 canines in the front on the bottom. The canines are not pointed but look like incisors. There are no incisors on the top; instead cattle have a dental pad. Cattle have 6 premolars and 6 molars on both top and bottom jaws for a total of 24 molars. In addition, there is a large gap between the incisors and molars. This configuration allows cattle to harvest and chew a large amount of fibrous feed.

Because their teeth are primarily for grinding, cattle use their tongues to grasp or gather grass and then pinch it off between their incisors and dental pad. Since they lack upper incisors, cattle cannot bite off grass very well, and they are inefficient at grazing closely. The inside of the cheeks and palate are rough which helps hold feed in while cattle chew with a side to side motion.

In addition to reducing the size of feed particles, the mouth aids in digestion by adding saliva to the feed. Cows will produce 20-35 gallons of saliva a day. The saliva helps moisten the feed. Saliva also contains sodium bicarbonate to keep the rumen at the proper neutral pH (6.5-7.2) for good microbial growth. Much of the water contained in saliva is recycled by the cow.

Stomach

The four compartments of the cattle stomach are the rumen, reticulum, omasum, and abomasum. The rumen is the largest compartment, and it contains billions of bacteria, protozoa, molds, and yeasts. These microorganisms live in a symbiotic manner with the cow, and they are the reason cattle can eat and digest large amounts of roughage. The rumen microorganisms are adaptable enough that cattle can digest a large variety of feeds from grass, hay, and corn to brewer's grains, corn stalks, silage, and even urea.

The bacteria and protozoa do most of the digestion of feeds for the cow. This is a tremendous factory. There are 25 to 50 billion bacteria and 200 to 500 thousand protozoa in every milliliter of rumen fluid (about 0.06 ounces). The microorganisms digest the plant fiber and produce volatile fatty acids. These fatty acids are absorbed directly through the rumen wall and supply 60 to 80 % of the energy needed by the cow. In addition to energy, the microorganisms produce protein including essential amino acids from the protein and nitrogen the cow ingests. Because the microbes can use nitrogen to make protein, cows can eat urea and other sources of non-protein nitrogen that would kill non-ruminants. The microbes also make vitamins B and C.

The reticulum, with its honeycomb like lining, is a compartment of the stomach that is involved with rumination. It also acts as a trap for foreign objects ingested by the cow. It is not unusual to find rocks, nails, and pieces of wire and metal in the reticulum of cattle. If wire or metal punctures the side of the reticulum, it can cause "hardware disease." Hardware disease is actually an irritation or infection of the diaphragm, heart or lungs. It is hard to treat, but can be prevented by keeping metal trash out of pastures. Specially shaped magnets can be administered to cows to decrease the possibility that ingested metal will pierce the digestive tract. These magnets stay in the reticulum for the life of the animal.

When cattle ruminate, or chew their cud, they are regurgitating a bolus of incompletely chewed feed. In order for the microbes to digest fiber rapidly and efficiently it must be in small pieces, so cattle re-chew their food several times. Cows also eructate or belch giving off carbon dioxide and methane. When cows "lose their cud" or stop ruminating, it is an indication that they have a digestive upset, and their rumen is not functioning properly.

Bloat is another condition that occurs when cows can't eructate. It is caused by a rapid change in feed or overeating grain (gaseous bloat) or grazing pure stands of clover or alfalfa (frothy bloat). Gaseous bloat is a result of improper digestion or fermentation of grain. It is treated by passing a tube into the rumen or using a trocar and cannula to make an external opening in the rumen to release the gas pressure. The procedure may have to be repeated. Frothy bloat is a result of surfactants in legumes causing gas to be trapped in a bubbly foam. Large amounts of mineral oil must be forced into the rumen via a tube to break up the bubbles as a treatment for frothy bloat. Bloat must be treated quickly as the increased rumen size and pressure interferes with normal breathing.

The incidence of bloat in cattle grazing legumes can be reduced by maintaining at least 50% of the stand as grass. Also, cattle should not be turned out onto a pasture with a high percentage of legumes when cattle are hungry or the pasture is wet. Once cattle are adapted to legume/grass pastures, they can graze it even when wet. A final option is to use "bloat guard" blocks which contain poloxolene.

Although rumen microbes can digest a great variety of different feeds, they are very sensitive to drastic changes in feeds. Some groups of microbes are better at digesting fiber (forages), whereas others are better at digesting starch (grains). Changing rapidly from a forage-based diet to a grain-based diet causes millions of fiber-digesting microbes to die-off as they cannot digest the starch, and there are too few starch-digesting microbes to use the grain so the grain sours in the rumen. As a result, rumen pH decreases, the rumen stops working, and the animal becomes ill. In severe cases, cattle can develop acidosis and founder or die.

The omasum is also known as "the book" or many piles because of its many leaf-like folds. It functions as the gateway to the abomasum, filtering large particles back to the reticulorumen and allowing fine particles and fluid to be passed to the abomasum. Though the complete function of this compartment is unknown, it does aid in water resorption and recycling of buffers for the saliva. The omasum may also absorb some volatile fatty acids.

The abomasum is also known as the "true stomach." It functions much like the human stomach producing acid and some enzymes to start protein digestion. Animals that go off feed or have acidosis can develop a displaced abomasum or "twisted stomach." The abomasum will actually float out of place and become torsioned stopping the flow of digesta. Surgery is the only cure for a displaced abomasum. Although displaced abomasum is more common in dairy cattle than beef cattle, producers should be aware of the possibility of this problem in cattle that have had severe digestive upsets.

Lower Digestive Tract

The rest of digestion is performed in the small intestine and large intestine much as it is in humans and other mammals. Digesta that leaves the rumen and enters the lower digestive tract includes some microbes and undigested fiber, as well as protein and some sugars produced by the microbes. By-pass protein, fat, and carbohydrates also enter the lower digestive tract. By-pass protein, fat, and carbohydrates are nutrients that cannot be digested in the rumen but may be digested in the abomasum and small intestine.

Enzymes to digest proteins, sugars, and starch flow into the small intestine from the pancreas, while the gall bladder produces bile to help digest fats. The small intestine also produces some enzymes to aid in digestion, but its major function is absorption of digested nutrients. Except for the volatile fatty acids, most of the nutrients are absorbed in the small intestine including protein, starch, fats, minerals and vitamins.

Water is primarily absorbed in the large intestine. Undigested feed, some excess water, and some metabolic wastes leave the large intestine as fecal material. The consistency of manure is an indicator of animal health and is dependent on water, fiber, and protein content of the feed. For example, cattle on lush spring forage will have profuse watery, greenish colored manure, whereas animals on a hay diet will have firm manure that is dark in color. Animals should produce manure that is indicative of the diet they are receiving. If not, it may indicate a digestive upset or disease. Light colored manure, manure tinged with blood, and watery manure (when on a dry diet) are not normal situations. Manure should not smell putrid or rancid. Producers should recognize changes in manure that indicate problems.

Feeding Management

Although proper nutrition and cattle health goes beyond taking care of the rumen microbes, reducing digestive problems and promoting a rumen with a healthy microbe population can prevent many serious problems in cattle. The following are some rules for maintaining rumen health:

- Provide a diet that meets the energy, protein and mineral requirements of the animal.
- Make sure water is clean and available.
- Pay attention to fiber levels in diets; a fiber level between 30% and 70% is preferred.
- Switch from high fiber diets to high grain diets slowly; change should occur gradually over days or weeks.
- Diets should contain 5% or less fat.
- Changing from a high grain diet to a high fiber diet will generally NOT cause digestive upsets, but will reduce performance.
- Do not move hungry or newly received cattle into pastures containing a high percentage (>25%) of legumes.
- Do not introduce cattle to high percentage legume pastures when pastures are wet.
- Supply cattle grazing pasture containing >50% legumes with ionophores or poloxolene.
- Monitor rumination and fecal output to aid in early detection of digestive problems.

Animal Nutrition-based Technologies and Practices

Nutrition is the foundation of a livestock production system and proper nutrition is imperative for achieving high and sustained livestock productivity. The success of animal reproduction and health programmes rests on proper nutrition. During the last four decades a number of animal-nutrition-based technologies and practices have been developed and applied both on-station and on-farm in developing countries, with varying degrees of success. Some technologies have produced profound beneficial effects and have been widely used; while others have shown potential on research stations but have not been taken up by farmers. Other nutritional strategies produced benefits to farmers so long as they were supported by a donor-funded project, but their use could not be sustained after the project concluded.

Change in Animal Nutrition Research Paradigm

Crop-livestock mixed farming prevails in India and the majority of rural families own livestock. Livestock production in countries like India has characteristic features such as:

- Prevalence of crop-livestock mixed farming system;
- High diversity and multi-functionality;
- Dependence of a large number of rural families on livestock for their livelihood;
- Increasing demand/prices of livestock products provide an opportunity for extra income;
- Ruminants dominate livestock production; and
- Feeding is based on 'Low External-Input System' utilising crop residues and by-products.

Livestock development provides a pathway out of poverty for the underprivileged families, however some constraints have to be overcome. Feeding of livestock is one such constraint and a challenge for the research system in India since the contribution of research to livestock development and particularly for the resource-poor farmers, is marginal.

Commonly reported observations on feeding of livestock focus mainly on two aspects: the majority of producers follow traditional feeding practices and they resist change/adoption of scientific recommendations and technologies. These observations imply that the fault lies with the livestock producer while the research system has done its job. There is hardly any attempt made to understand the reasons for non-adoption which raises an important question as to why the research system is not sufficiently concerned with the utility of outputs/ products of research. This situation indicates that there is probably a mismatch between the prevailing livestock production systems (that are not likely to change in the near future) and research systems (resistant to change). Some characteristics bearing on the relationship between these two systems are discussed below.

Whole-farm approach of the farmers as against sectoral approach of the research system. Livelihood and farming systems of the underprivileged are complex and livestock production is a crucial component. Most farmers are concerned about the output of the whole farm and not of livestock alone while animal nutrition research usually focuses only on the animal. As the sub-systems influence each other, changes in any sub-system tend to impact the whole farm.

Reductionist approach of the research system. Studies on production of greenhouse gases (GHG) by ruminants is a good example of the reductionist approach. It is only the enteric GHG production that is estimated and no attempt is made to get a total picture based on production as well as savings of GHG due to the feeding systems of ruminants applied. Mishra and Dixit studied the total picture and reported that GHG produced by ruminants are balanced by savings due to the feeding systems adopted in India. Use of crop residues and agricultural by-products saves GHG that would have been emitted in producing grains and making grain-based livestock feeds. These uses also save the GHG that would have been produced by burning of crop residues. Recommending low fibre/high energy rations to reduce methane production is also inappropriate for Indian conditions.

Methane production by different types of animal is compared on the basis of amount of methane per unit of milk produced, which is not appropriate since the majority of cattle and buffalo, in a country like India, are not dairy type. A different approach that also accounts for other useful services, functions and products for farmers is needed for comparing GHG production. There is a need to consider farmers' perceived value of these services and products, however, as these are not quantifiable by conventional methods, a social-cost accounting approach may have to be used.

Livestock production in India is highly internalised. The system is mainly low external-input oriented and thus tends to optimise the use of available resources (human as well as material). Livestock are mostly grazed and supplemented by farm produce (crop residues, by-products, tree fodders), hence the resistance to adopt new feeding systems or technologies that require purchase of material or hire of labour.

The Indian Council of Agriculture Research (ICAR) led the Indo-Dutch project on Bio-conversion of crop residues which was implemented in the 90s through a network of centres in different regions. This project demonstrated the use of systems and participatory approaches to understand the farming system in general, with in-depth understanding of livestock sub-systems in various agro-ecozones of India. The project showed a way of planning animal nutrition research based on the understanding of systems and the need to define conditions under which each recommendation/ technology would work.

While the role of women in feeding of livestock is well known the research system has not given it due consideration. Studies on the involvement of women in livestock production show women to have good knowledge of feeding habits of animals and of local feed resources, and that women take decisions related to feeding of animals. However, the research system rarely involves women or considers their knowledge and views in planning studies, conducting trials or assessing the research outputs.

Some of the conventional recommendations for livestock feeding need reconsideration. Rangnekar, Schiere and Rao have discussed recommendations e.g. feeding milk or milk replacer to calves and kids at the rate of 10 percent of body weight, and feeding of greens equivalent to 33 percent of total roughage dry matter. Feeding of milk is probably recommended for ideal gain in body weight, however, hardly any small farmer can afford to offer that much milk to calves. Green fodder production is taken up only on small plots and its availability will always be limited. Most recommendations suit high external-input systems, are aimed at maximising production (of a particular commodity) and are not appropriate for smallholders. Moreover, there is variation in systems and resources between and within regions of the country, thus making uniform recommendations inappropriate. Critical assessment of reasons for non-adoption of recommendations would provide useful guidance for the future.

Factors Influencing Famers to Adopt New Practices and Technologies

The decision to change from traditional feeding systems and to adopt new practices and technologies is a complex process. Contrary to common belief the decisions are not based on economic benefit considerations alone; several more factors come into play. A good understanding of the decision-making processes is useful in the selection of appropriate research outputs for recommendation.

The factors usually considered by farmers while deciding change of practices or adoption of technology were listed and discussed by Rao *et al.* These factors range from relative advantage, observable results, divisibility, simplicity/complexity and initial cost to compatibility of the recommendations with the farming and social systems. Another factor crucial in deciding adoption is the 'relevance to constraints/problems' faced by livestock owners. Hence, only outputs of research (feeding recommendations or technologies) that are technically sound, economically beneficial, socially acceptable, suit the prevailing farming system and solve current problems stand a chance of adoption.

In a study of the role of women in urban livestock production, in western India, it was observed that most of the commercial dairy producers in urban areas continue to make feed mixtures and feed animals in the traditional manner contrary to the reports that educated, commercially-oriented and urban livestock producers are early adopters. Multidisciplinary research may provide better understanding of the reasons for continuing traditional practices of feeding livestock by these urban producers.

Roy and Rangnekar conducted a participatory study to understand the reasons for adoption of urea treatment of cereal straws in the operational area of Vaishali cooperative milk union in the state of Bihar, in eastern India. The farmers reported three main benefits of urea treatment, namely prevention of spoilage of straw due to unseasonal rain, more straw could be stored in the bins and marginal increase in milk production. Other factors contributing to adoption of the technology were: initiative of dairy cooperative in demonstration of the technology, availability of straw and water, shortage of legumes in summer and high level of milk production of the cows. According to recent reports, more than 4, 000 farmers adopted the technology in 2010. It can be concluded that urea treatment of straws fitted well with the straw handling system in Vaishali area, and benefitted the whole farm operation of the farmers (not milk production alone).

Making Animal Nutrition Research Resource-Poor Friendly

To be effective, development needs research, and without linkage with development, research remains academic. There is a need to promote pro-poor approaches in livestock research, similar to that done by the FAO, for livestock development. Hazel and Haddad recommend that public sector research in developing countries should give priority to developing a strategy for pro-poor research through strategic planning. Moreover, to develop pro-poor research there should be good understanding of livelihood systems and adoption of a 'participatory approach involving women'. Developing technologies and prioritising research areas on the basis of agro-ecological and socio-economic conditions, at national as well as regional levels, is recommended.

The participatory research should not be 'scientist led' but all the stakeholders including farmers should be involved from planning to the stage of interpreting the results The study of traditional feeding practices and perceptions of the livestock owners, especially the women, about feeding of livestock is suggested to 'understand why farmers do what they do'. The large network of Krishi Vigyan Kendras (technology transfer centres in India), universities, research institutes and centres of the ICAR in India should enable adoption of such approaches without much difficulty. However, there would be need to orient animal nutrition research workers to develop close linkage/interaction with extension system and farmers.

To further explain the nature of change proposed in animal nutrition research, the case of studies on GHGs in ruminants is taken as an example and suggestions are given below:

- Estimate total GHG production by local ruminants under traditional feeding regime;
- Undertake balance study of GHG using the approach and social cost accounting method suggested by Mishra and Dixit to get a holistic picture under different production systems; and
- Screen various feed materials including tree leaves and pods, in different regions, for effect on GHG production in ruminants. Identify low-cost locally-available materials that reduce GHG and can be used by an average livestock owner. Used by a large number of farmers would have impact, even with a marginal decrease in methane production.

There are two new initiatives developing in India that are worth considering. One is in the crop sector known as "System of Rice Intensification or SRI" and based on participatory research and technology development. It is a joint initiative of farmers, NGOs and scientists and is fast emerging as an alternative to conventional high-input system of rice cultivation and is being tried with other crops. The other initiative is "Knowledge Swaraj" taken by scientists and a civil society organisation to renew the relevance of traditional knowledge with the objective of guiding the use of science and technology for development of India.

Urea-molasses-multinutrient Blocks/Licks

A comprehensive survey conducted to assess the nutritional status of dairy animals in six different agro-climatic zones of Punjab state of India revealed that animals of all zones, except that of central plain zone (CPZ) and western plain zone (WPZ) were being offered highly-imbalanced diets, deficient in protein and/or energy or minerals. The diets offered were unable to meet the nutrient requirements, as also indicated by low milk urea-N (4.67 to 9.45 mg/dl). The proportion of roughage in the complete feed was highest (86.9 percent) in undulating plain zone (UPZ) and lowest (71.4 percent) in western zone (WZ), indicating that the rations of dairy animals under field condition were roughage based. Only 19.7 percent of farmers offered mineral mixture (MM) in CPZ while only 2.5 percent in sub-mountainous undulating zone (SMUZ). Forty-eight percent of farmers in CPZ and 19.7 percent of farmers in WZ supplemented the diet with common salt. The deficiency in minerals and salt was reflected in the reproductive problems of animals in these areas. The highest cases of repeat breeding (anoestrus) were recorded in CPZ and WPZ (52.0 and 51.5 percent respectively) and the lowest in WZ and FPBZ (28 and 34 percent respectively). It was clear from the survey that feeding of a balanced diet (with respect to energy, protein and minerals) must be advocated under field conditions. Wheat/ rice/ragi straws, and maize stover/ millet stalks constitute the bulk of dry matter (DM) consumed under field conditions. Supplementing such diets with UML/UMMB would be beneficial to ruminants.

Development of the Technology

Several processes have been developed to slow down the release of ammonia from urea in the rumen. Uromol – a Maillard product formed by heating urea and molasses (1:3 ratio w/w; heating for 20–30 minutes) was developed by the Department of Animal Nutrition, Punjab Agricultural University, Ludhiana. To avoid solidification of Uromol at room temperature, it was mixed (whilst hot) with four parts of wheat or rice bran. The ensuing mixture was named Urobran. Rate or rumen release of ammonia from uromol/ urobran was slow and consistent leading to better nitrogen utilization by rumen microbes compared to urea-molasses fed without heat treatment. Urobran was shown to be suitable as a complete replacer for groundnut or mustard cake in a concentrate mixture. Uromol-straw combinations could be safely used as a basal ration for ruminants provided adequate energy was available as digestible organic matter. To reduce fuel costs and make the process more farmer-friendly, the heating was dispensed with and urea-molasses-multi-nutrient blocks (UMMBs) were prepared by the cold process. The heat generated by mixing calcium oxide with urea and molasses helped bind the urea and molasses via the Maillard reaction. Nutrient utilisation in buffalo calves was comparable for UMMBs prepared by hot and cold process.

Ummb Preparation and Palatability

UMMB is a convenient and inexpensive method of providing a range of nutrients to animals. It can improve the digestion of low-quality roughages by satisfying the requirement of rumen microorganisms, creating a better environment for fermentation and increasing production of microbial protein and volatile fatty acids. Urea, after hydrolysis to ammonia in the rumen, provides a nitrogen source for the rumen microbes, while molasses acts as a source of readily-fermentable energy.

The conventional UMMB weighing 3 kg contains (g): molasses 900, urea 300, mustard cake 300, de-oiled rice bran 300, wheat flour 450, mineral mixture 450, calcium oxide 120, salt 120 and guar gum 60. The required quantity of molasses and urea are weighed and mixed in a 25 kg capacity iron pan. The guar gum is added to the urea-molasses mixture as a binder. A premix of other ingredients is prepared; calcium oxide is the last ingredient to be added to this premix in the iron pan with rapid stirring. Two to eight UMMBs may be prepared at a time, either in a manually operated (for small/ marginal/landless dairy farmers) or electrically powered (for commercial production) block-making machine.

On average an animal licks 500–700 g UMMB/day, but some animals may lick as much as 1.25 kg/day without any adverse effect on their health.

The hardness of the UMMB affects its intake; this aspect acquires greater importance when medicated blocks are used for drug delivery, since dosages are based on intake of such blocks. Animals do not seem to like licking very hard blocks, while soft blocks are chewed rather than licked. Blocks that show no finger impression on pressing, are considered to be of desirable hardness.

Ummb Containing Non-Conventional Feed Resources

The blocks may also be prepared by using non-conventional feed resources. Iso-nitrogenous and iso-energetic licks were prepared by replacing: a) wheat flour with waste bread (an excellent source of by-pass starch containing 11–12 percent crude protein) and b) oiled mustard cake with tomato pomace (20–22 percent C P, 10–11 percent ether extract and a 'good' amount of lycopene, an antioxidant and pigment that gives colour to meat). The combination of alternate energy and protein sources improved the utilisation of nutrients. The metabolisable energy (ME) content of UMMBs containing non-conventional feed resources varied from 5.71 to 6.03 MJ/kg DM. *In vitro* gas production studies also revealed similar efficiency of nutrient utilisation from various UMMBs. Adult male Murrah buffalo offered 1 kg concentrate, 5 kg green fodder and 9 kg wheat straw daily, supplemented with ad libitum UMMB consumed 1.08 kg of conventional block compared with 1.32–1.84 kg experimental blocks. The UMMB supplementation improved intake of wheat straw, CP digestibility and N-retention. The use of such wastes can reduce the cost of producing UMMB because tomato pomace is available free, whilst waste bread is available at Rupees 7/kg compared with wheat flour at Rupees 15/kg .

The cost of molasses can be reduced by using spent sugar syrup available at nominal charges or free from the 'Muraba' (fruits preserved in sugar syrup) manufacturing industry. Bakshi, Wadhwa and Parmar (unpublished) used spent sugar syrup in place of molasses for manufacturing UMMB. The results were comparable with those recorded for UMMB containing molasses.

Impact of Ummb on Performance of Animals

Due to mineral deficiencies, pica is a common problem in almost all the animals in arid regions. Pica was effectively reduced by UMMB supplementation. In some cases, animals suffering from haemoglobinurea due to phosphorus deficiency recovered when supplemented with UMMB.

Farmers reported animals to be better general body condition, with glossy coats and healthier appearance. Cows that had not shown oestrus signs for a long time (presumably due to inadequate nutrition) resumed cycling when given mineral-rich blocks. Increases in milk production due to UMMB supplementation have generated additional income whilst improving reproductive performance, leading to more calves. These improvements have undoubtedly improved the socio-economic status of farmers.

With UMMB supplementation, first service conception rate was improved (from 41.4 to 56.7 percent), while services per conception declined (from 2.54 to 1.88). UMMB supplementation also reduced culling rate due to infertility (from 8.6 to 3.2 percent). Out of 64 observations on calving intervals, 42.2 percent were below 360 days, compared with 50.0 percent after UMMB supplementation. UMMB supplementary feeding during the pre-partum period improved post-partum reproduction in terms of days to first oestrus (from 48 to 34 days) and conception rates (from 0 to 30 percent).

The response to supplementary feeding with UMMB was more pronounced in buffalo kept in rural areas by marginal farmers than in those kept on organised farms. This was due to the differences in basal diet of the systems. UMMB supplementation of early-calved buffalo induced a higher proportion exhibiting oestrus during the first 50 days post-partum compared to controls. Pre-partum UMMB supplementation improved milk yield, and peak milk yield was maintained longer during the post-partum period. UMMB supplementation for 30 days in delayed-puberty buffalo heifers induced oestrus in 33 percent of cases in summer, and in 93 percent of cases in winter. Similarly in anoestrous buffalo, UMMB supplementation induced ovarian activity in 40 percent of buffalo during the summer season and in 90 percent during the winter season. In addition, UMMB supplementation was shown to increase the effect of pregnant mare serum gonadotropin used to induce oestrus in anoestrous and delayed-puberty buffalo.

Medicated Ummb

Supplementary feeding of recently-calved buffalo using UMMB medicated with Replanta (an indigenous herbal preparation) decreased the time for shedding of placenta (5.75 vs. 4.40 hours) and the days to first post-partum oestrus (42.5 vs. 36) for treated and control group, respectively. UMMB containing anthelmintic has been successfully used for controlling nematode parasites.

Adoption of The Technology

Mineral mixture prepared as per Bureau of Indian Standards specifications is being used for manufacturing UMMBs by the cold process and also using conventional feed ingredients in an electric block-making machine. Using the machine, eight blocks are made at a time and this is achieved within three to five minutes. Currently, UMMBs are prepared under a Revolving Fund Project and a sum of Rupees 4 million has been provided by our university for the project. UMMBs are sold at Rupees 70 per 3 kg block (1 US$ = *ca* 45 Indian Rupees). The UMMBs are sold all year round, from the premises of our department. UMMBs are also sold in the livestock shows organised by the University twice-a-year, and sold too in the zonal as well as regional cattle-judging competitions organised by the State Animal Husbandry Department. In addition, the Krishi Vigyan Kendras (an institution under Indian Council of Agriculture Research, responsible for technology transfer) purchase the UMMBs from our department and sell them to farmers. During the last three years the sale of the UMMBs has increased from 2 660 to 6 530/annum. Farmers are keen to buy the blocks and feed them to animals. Some cattle-feed manufacturers and government agencies, e.g. Markfed and Milkfed have taken up these technologies. The future of these technologies would appear to be very bright.

Straw-based Densified Complete Feed Block Technology

Most of the countries in the tropical region face an acute feed shortage. Crop residues, especially cereal straws are the staple feed available in this region for ruminant feeding. These fibrous and bulky feeds are of low palatability and digestibility. Also crop residues have low available energy, protein and mineral contents. Apart from feed shortage, India is also facing regional feed disparity, with some states like Punjab, Haryana and Western UP growing sufficient green fodder and being surplus in straws, while other states lack in the supply of greens and dry roughages. The bulky nature of crop residues makes their transportation costly and difficult. Straws worth millions of dollars are therefore burnt in the fields; this not only wastes a feed resource, but also causes environmental pollution and soil degradation. Therefore there is need to improve the management of feed resources in India. One improvement would be using specially designed balers for collecting straw from fields and then subjecting straw to processing technologies for the commercial production of balanced animal feeds.

Technology of Straw-based Densified Complete Feed Blocks

A novel concept in nutrient delivery system for ruminants in the tropics. The technology for making straw-based Densified Total Mixed Ration (DTMR) or Densified Complete Feed Blocks (DCFB) is a novel and revolutionary approach, which provides an excellent opportunity for feed manufacturers and entrepreneurs to remove the regional disparities in feed availability and also to supply balanced feeds on a large scale, to dairy and other livestock farmers. This approach could overcome the problem of feeding animals during natural calamities such as floods and droughts, which are increasingly common in the tropics. The technology has potential for use in other tropical countries where farmers find it difficult to compute balanced feeds. These difficulties result in below-optimum levels of livestock production. Each DCFB is a total ration for a cow or buffalo and supplies all the major and minor nutrients, including micronutrients required by the animal.

The two major components of straw-based DTMR/DCFB are roughage and concentrate, added in different ratios, depending on the level of production, stage of lactation and physiological state of the animal. The roughage part is generally crop residue e.g. straws, stovers and sugar cane tops. Sometimes gram straw, sugar cane bagasse and groundnut haulm are also used as part of the roughage component. In hilly areas, even non-conventional roughage sources like dry forest grasses and tree leaves have been used in place of crop residues. The proportion of straw and concentrate in the block varies with the type of animal to be fed. The ingredients of the concentrate mixture are oilcakes (as protein source), molasses, grains, grain products/by-products (as energy sources) and supplements like by-pass protein/by-pass fat, to enhance the direct supply of amino acids and fatty acids to the host animal. The third component provides strategic and catalytic supplements, e.g. vitamins, minerals, bentonite (toxin-binder), probiotics, antioxidants and herbal extracts. The role of this component in the feed block is to enhance the productive and reproductive efficiency of the animal and to enhance the immuno-protective ability to keep the animal disease-free.

Method of manufacturing feed blocks. The making of feed blocks requires appropriate processing. Blocks can be manufactured on a large scale in a factory. The concentrate mixture is prepared separately and then mixed with straw (in a specified ratio) in a specially-designed mixer to ensure that the two components, which vary greatly in their densities, are adequately mixed. The mixed feed is compacted in a specially-designed hydraulic press.

With the help of a binder, the process of densification results in concentrate particulate-matter becoming physically attached to fibrous straw particles. The whole process brings uniformity to the feed, prevents selection of ingredients by the animal, increases palatability and minimises the feed wastage. A few agencies have taken the initiative of developing and refining the feed-block technology in India. Fuel-efficient and labour-efficient models of machines for feed-block production have been developed in India. Results of on-farm feeding trials with feed blocks, conducted recently on buffalo in rural areas around Karnal, showed a significant increase in milk production (10–15 percent) compared with normal feeding practised by farmers in the villages. Although feed cost increased by 30 percent with block feeding, the profit earned was approximately 20 percent higher.

Experiences of Applying Technology/Practice by Farmers

The DCFB technology offers a variety of benefits, both to the farmer as well as to the feed entrepreneur. The following advantages of this technology have been identified.

A balanced ration for ruminants. In India, except at some organised farms, the practice of feeding a balanced ration is almost negligible. The DTMR/DCFB is a complete, balanced feed. By feeding a balanced feed with optimum nutrients, one can expect improved nutrient utilisation, resulting in optimum productive and reproductive performance from the animal.

An efficient nutrient-delivery system. The feeding is simple and hassle-free. The animal is not able to select ingredients. In the system where ingredients are fed separately, the animal picks up the more digestible/ palatable part first, followed by the less-palatable and lower-quality component of the feed.

Time and labour saving. The labour requirement of feeding is reduced by 30–40 percent. Twenty animals can be fed in about 10 minutes as against 2 hours required for feeding the same number of animals in a conventional feeding system. DTMR/DCFB could also be a clear advantage to women, especially in hilly areas, where women generally look after the feeding and management of dairy animals. In these areas, women spend most of their time in drudgery having to cut, collect and transport large loads of grasses and tree forages from the forest areas.

Cheaper and easier transportation. Densification of the straw caused a 3-fold reduction in its bulkiness. Accordingly, less space is required to store the feed, especially straws. Because DCFB occupy less space/volume, almost 3 times more feed (by weight) can be accommodated per load, making transport of crop residue-based feeds much easier and cheaper. Blocks are also easier to handle during storage.

Can check environmental pollution. The emission of methane gas (a greenhouse gas) from ruminants can be reduced on feeding densified feed blocks. The feeding of a balanced diet reduces the methane emission by 10 to 15 percent. There is also less dust pollution when blocks rather than loose straws are transported in over-loaded trucks. The latter are also a traffic hazard. In north-western parts of India, straws worth millions of dollars are burnt in the field after the harvest. Residual straw converted into feed blocks can be efficiently utilised and environmental pollution can also be reduced. Burning of straw in the field also lowers soil fertility.

Improved productive and reproductive efficiency. Feeding DCFB can increase growth rate of calves by 25 to 30 percent and milk yield by 10 to15 percent. Also the milk yield persists longer, causing an increase in total lactation yield. The absence of dietary fluctuations results in a relatively stable microbial-ecology in the rumen which increases its efficiency. Feeding CFB results in earlier maturing of animals. This not only lowers the cost of rearing, but also reduces age at first calving. It also provides regularity to subsequent calvings and increases life-time production. The optimum supply of nutrients, including micronutrients has a positive effect on health which keeps the animal free from many reproductive problems e.g. late maturity, anoestrus and repeat breeding.

Storage of bulky feeds possible. With the availability of feed-block technology, it is possible to set up feed/fodder banks in feed-deficit areas, because of the ease of handling, transport and storage blocks. Feed block technology can also be beneficial during natural calamities, e.g. floods and drought. The blocks can be air-lifted to the remotest places to avert disasters.

Vehicle for feed additives/pharma/nutraceuticals. DCFB offers a delivery system for specific nutrients, nutraceuitical and medicines e.g. anthelmintics to control parasites in livestock.

Processed Crop Residue-based Complete Diets

The efficient utilisation of crop residues and agro-industrial by-products as animal feed has assumed importance in India due to shortages of dry roughages, concentrates and green fodder which were estimated to be 19,

62 and 45 percent respectively (Anonymous, 2008). Due to cultivation of food and commercial crops, enormous quantities of different types of crop residues are produced as a renewable resource every year. Most of them are being wasted or used as fuel in some villages. Crop residues are rich in fibre and low in nitrogen, minerals and vitamins. Hence, their palatability is low and therefore crop residues cannot form a sole ration for livestock. Any processing method that improves the nutrient availability from crop residues will assist in bridging the gap between availability and requirement of roughages.

The concept of complete feeds for ruminants is relatively recent in India. In this system, all feed ingredients inclusive of roughages are proportioned, processed and mixed into a uniform blend. The product is fed as a sole source of nutrients. This system ensures an adequate supply of balanced nutrients to the animal, controls the ratio of concentrate to roughage, helps in improving utilisation of low-grade fibrous crop residues and reduces feed wastage and feeding cost. The system also promotes feed intake and avoids refusal of the unpalatable portions of feedstuffs. Such rations also reduce eating and rumination time and increase resting time.

An even intake of feed is associated with less fluctuation in rumen-release of ammonia so that non-protein nitrogen may be more efficiently utilised. This system gives good scope for increasing the use of home-grown fibrous crop residues and by-products. The level of inclusion of different crop residues in complete diets of ruminants is shown in Table 1.

Table 1: Inclusion levels of different crop residues in complete diets of ruminants

Crop residue	*percent in complete ration*
Sugar cane bagasse	20
Sorghum straw	20–46
Dry mixed grass	30–75
Sunflower straw	30–50
Sunflower heads	33–50
Wheat straw	50
Fallen teak leaves	17.5–70
Mango leaves	30–60
Rice straw	40–50
Cotton straw	45

The roughage portion varies from 30 to 70 percent depending upon the physiological status of animals to be fed. The roughage content of complete diets is 30–40 percent for high-yielding animals, 40–50 percent for growing animals and 60–70 percent for dry animals, whereas the protein content varies from 11 to 13 percent. The proportion of other ingredients varies accordingly. These diets can be processed into mash or pellets. For pelleting of mash feed, traditional steam pelleting or recently developed expander-extruder technology can be adopted.

Processing of Complete Diets

Mash. The different ingredients of the diet are proportioned as per formula into 100 kg batches using appropriate scales. A hammer mill with 8 mm sieve is used for grinding the ingredients. Through screw conveyor and bucket elevator the ground ingredients are discharged into the hopper fixed on top of the mixer. The ingredients that do not require mixing are directly added to the mixer. Molasses is pumped from the storage tank to a heating tank, where it is heated to 70 °C. The heated molasses is sent to the mixer through a dosage tank, as per formula. The micro-ingredients (vitamins and minerals) are made into a premix by mixing them with ground grain or bran and the required quantity of premix is added to the mixer directly. Later, all ingredients are mixed for about 10 minutes and then collected in sacks and stored.

Pelleting. For pelleting, the mash from the mixer is dropped into bucket elevator and lifted and conveyed into a hopper over the pellet mill. The feed, in mash form, is conveyed from the hopper into the conditioning chamber of the pellet-mill through a screw conveyor. A wheel-valve controls the rate of flow of the feed into the conditioning chamber. The steam produced from a boiler attached to the pellet-mill is supplied into the conditioning chamber of the pellet-mill. The required quantity of steam at 97–98 °C is supplied into the conditioning chamber through a control valve. Conditioned mash at 90–92 °C and 16–17 percent moisture is conveyed to the pellet mill and extruded through a ring die with 9 mm holes. The pellets of 9 mm diameter at 83–85 °C and 14–15 percent moisture are dropped from the pellet mill into a vertical cooler, fixed below the pellet mill. The cooled pellets are collected into sacs.

Expander-extruder processing. This is a system which combines the features of expanding (application of moisture, pressure and temperature to gelatinise the starch portion) and extruding (pressing the feed through constrictions under pressure). The mash containing 12–13 percent moisture and at room temperature is reconstituted with the required quantity of water to get 17–18 percent moisture in the mixer and then sent to the hopper above the expander-extruder from which it passes through a screw in which it attains 90–95°C by the time it comes out of the die openings. Otherwise, the mash without reconstitution can be sent to hopper and steam is added to get the required moisture while the feed is passing through the screw of expander-extruder. The pellets coming out of the expander-extruder are cooled and collected into sacks.

Experiences of Applying Technology in the Field

To demonstrate the advantages of the complete diet for economical milk and meat production, three complete diets containing a) cotton stalks, b) maize cobs, and c) sunflower heads as sole roughage sources (28.5 percent in dairy ration and 40 percent in sheep and goat ration) were fed to Murrah buffalo/sheep and goats owned by farmers in the state of Andhra Pradesh, India.

Milk production studies. The diets maintained 6–8 litres/day of milk yield in lactating Murrah buffalo. Dry matter intake was 28.2, 26.7 and 15.0 percent less on complete diets containing cotton stalks, maize cobs and sunflower heads, respectively, compared with the conventional diet. Average milk yield increased by 11.3, 11.5, 23.5 percent and cost of feed/kg milk yield reduced by 21.5, 23.4 and 25.2 percent on feeding complete diets containing cotton stalks, maize cobs and sunflower heads respectively, compared with the conventional diets. The results indicated that cotton stalks, maize cobs and sunflower heads can be incorporated in complete diets as sole sources of roughage without any adverse effect on milk production in lactating Murrah buffalo. The milk yield was 16.7 and 11.4 litres/day on feed blocks produced from premium and lower-quality sorghum stover, respectively indicating the feasibility of medium levels of milk production on crop-residue-based complete feeds.

Meat production studies. The diets supported 87–108 g average daily gains (ADG) in Nellore ram lambs and 71 to 81 g ADG in local male kids in intensive-feeding system. The ADG was significantly ($P<0.05$) higher on complete diets compared with the respective conventional diet in both species. The dry matter intake/kg gain decreased by about 20 percent in sheep and 15 percent in goats while the cost of feed/kg gain decreased by 13 to 32 percent in lambs and 13–22 percent in kids compared with the conventional diet. Dressing percentage and meat: bone ratio were optimum and almost similar for complete and conventional diets in sheep and goats.

These results indicate that complete diets containing cotton stalks, maize cobs and sunflower heads can sustain optimum growth in lambs and kids under intensive system of feeding without any adverse effect on meat characteristics. In addition, enteric methane production can be reduced considerably by formulating a balanced complete-diet with different concentrate ingredients and crop residues. Gross energy lost as methane reduced ($P<0.01$) from 5.64 to 4.90 percent and from 5.85 to 5.14 percent in buffalo and cows, respectively. Further, with this system of feeding, migration of sheep and goats during scarcity periods can be avoided.

Based on the response of farmers in on-farm evaluations, a small-scale complete-feed processing unit capable of grinding all crop residues including cotton stalks, was developed under National Agricultural Technology Project (NATP) for use at village level. Twenty four units of small-scale complete-feed processing units were distributed to different research institutions and farmers in India. Two farmers established the same units on their own farms for feeding dairy animals and sheep in the state of Andhra Pradesh, India. This enabled the farmers to utilise their crop residues effectively.

Future of the Technology

Owing to human population-growth, increasing urbanisation and rising incomes, the demand for milk and meat is projected to double by 2020 in India. At the same time there will be increased requirement for food grains (cereals, pulses and oilseeds) which will result in decreased land for cultivation of fodders and in increased availability of crop residues. Ruminant livestock in developing countries, including India, will have to depend increasingly on crop residues to meet their nutrient requirements. To make the ruminant production system efficient, the crop residues will have to be processed.

Processing of Poor-quality Crop Residues as Livestock Feed

The key to improving the utility of PQCRs (straws, stalks and stovers) for ruminants is to overcome the barriers which limit their microbial fermentation in the rumen. The two major factors limiting microbial digestion in the rumen are high lignin and low-nitrogen contents. A 10 percent increase in cell-wall digestion would result in saving of 2 million tonnes of grain supplements and decrease manure solids by 2.8 million tonnes and increase animal productivity. A number of physical, chemical and biological treatments have been researched and developed worldwide in order to improve the utilization of PQCRs as feed for ruminants. However these technologies could not be sustained because of high cost of chemicals, depression in digestibility of PQCRs or problems in scaling-up or applying the technologies under field conditions.

Natural Fermentation of PQCRS with Urea

The recommended carbon to nitrogen ratio for the composting process of edible mushroom cultivation was maintained by providing 1.6 percent nitrogen in the substrate (wheat straw) through urea. The urea:wheat straw (3.5:96.5) mixtures, moistened to 70 percent and fermented by an open stacking method for 9 or 12 days (with or without turning at 3-day intervals) were evaluated on 6 adult buffalo. More than 85 percent of the added urea was hydrolysed by the 9th day and the straw that was not turned had the highest nutrient availability and nutritive value (NV). It was concluded that 9-day naturally-fermented urea wheat straw (FWS), daily supplemented with mineral mixture and vitamin A (or carotene) could meet both energy and protein requirements for maintenance of adult buffalo. The technology was not readily accepted by farmers because of the high moisture content of the treated straw. The technology was therefore modified using different levels of moisture (40, 50, 60 and 70 percent). The NV of straw fermented for 9 days at 40 percent moisture was comparable with the straw fermented at 70 percent moisture. The fermentation technology so-developed by using 40 percent moisture proved highly applicable under field conditions. The technology has universal application in improving the NV of PQCRs. The treatment can be done all year-round depending upon availability of the substrate e.g. wheat straw in April and paddy straw in November.

The natural fermentation technology involves a combination of treatments, i.e. chaffing straw (physical) and hydrolysis of urea and utilisation of ammonia-nitrogen by microbes for their proliferation, thereby enriching straw with micro-bial protein (biological), as indicated by significantly higher microbial population (bacteria, fungi and actinomycetes) in the stacks. Furthermore the rise in stack temperature to 55 °C (physical) facilitates ammonia from urea hydrolysis to penetrate cell walls and break the alkali-labile lignocellulose bonds (chemical). In the studies, the lignin content of the straw was not affected by natural fermentation because most of the microbes isolated from stacks (except Aspergillus fumigatus and Streptomyces sp.) were non-lignolytic. Pathogenic microbes such as Salmonella or E. coli were never detected in any of the stacks. The fermentation process could be hastened by using freshly-prepared fermented wheat straw as a natural crude inoculum (5 percent, DM basis). The 9-day inoculated straw preparation had significantly higher NV than un-inoculated fermented straw. The fermentation of inoculated straw could be terminated by the 6th day without affecting the NV and with minimum nutrient losses. Rumen fluid dilution and outflow rates, and the nutrient availability post-rumen in buffalo fed 6–9-day inoculated fermented straw were higher than when untreated or urea treated straw was included in the diets.

The feeding of FWS could sustain the growth of buffalo calves, and productive and reproductive performance of lactating buffalo even when supplemented with either a lesser quantity of concentrate or concentrate mixture with low-protein content. The incorporation of FWS with concentrate mixture containing non-protein nitrogen supplements (Uromol bran or deep-stacked poultry litter) in the ration of young calves had no adverse effect on productive performance. Use of naturally-fermented rice straw with urea, as livestock feed for calves improved the metabolisable energy content of feed by 15.5 percent and live weight gain by 33 percent. The use of fermented straw in livestock rations could spare about 70 percent of oil seed cakes.

Transfer of Technology

In a pilot project, 54 farmers from 21 villages of 4 districts of Punjab state were selected and were given repeated demonstration for upgrading wheat straw by natural fermentation containing urea (urea:wheat straw, 3.5:96.5) at 40 percent moisture by stacking for 9 days. Demonstrations with regards to gradual inclusion of fermented straw in the ration of livestock were also given regularly. Handouts in vernacular language were distributed periodically. Each selected farmstead was monitored for one year. The palatability of fermented straws was higher compared with untreated straws. Calves (above 6 months), heifers and lactating animals showed a remarkable improvement in physical appearance, health and milk production respectively.

Benefits of the Fermented Straw te Chnology

- Serves as a maintenance ration when supplemented with 25 g salt, 50 g mineral mixture and 2 kg green fodder;
- Improves the productive performance of calves above the age of 6 months as well as that of lactating animals, when supplemented with low-protein concentrate mixture;
- Spares oilseed cakes and improves the economics of dairy farming; and
- Controls environmental pollution

Major Constraints

Although the technology is exceedingly effective, it has not been adopted extensively. The main reason for low adoption is the male farmer being very busy in the day-to-day agricultural practices and being unable to spare sufficient time for dairy farming. This is one of the reasons for dairy farming being taken care of by the house lady or farm labourer.

Lack of mechanisation is the major constraint to widespread adoption of this technology at farm level. In northern India especially Punjab and Haryana, dairy farming is not a full-time occupation for the farming community, except commercial dairy farmers. Therefore, the farmer wants ready-made feed to be available at his/her doorsteps. If the technology is mechanised, it will not only save time and labour (availability of labour is declining), but will also make the production of treated straw economic.

Solid State Fermentation

Sixteen different lignolytic fungal strains were evaluated in our laboratory for upgrading the NV of straws with minimum pre-treatment, with or without exogenous nutrients. The proliferation of mycelia enriched the straw with microbial protein. The increase in *in sacco* dry matter (DM) and crude protein (CP) degradabilities of straws after 5-10 days of fermentation was associated with linear increases in DM loss in all the cases. *Heterobasidion annusum, Phellinus linteus, Phialophora hoffmannii and Cyathus stercoreus* (Bakshi *et al.*, 2010) or a blend of *Pycnoporus sanguineus and Oideodendron echinula* proved promising, provided fermentation was stopped before the onset of the reproductive phase to avoid high nutrient losses. In order to achieve the optimum DM and CP degradabilities with minimum nutrient losses, fermentation should be stopped within 6-8 days of spawning. Amongst the tested strains, only *Coprinus cinerieus* required nitrogen and phosphorus exogenously, and was found to be highly cellulolytic rather than lignolytic. Some workers have used urea-treated straw as substrate for *Coprinus* cultivation, but that adds to the cost and incurs loss in digestible nutrients.

Spent Straws

The wheat and rice straws available after harvesting edible mushrooms viz. *Agaricus bis-porus, Pleurotus ostreatus* or *Volvariella diplacea*, currently used as soil conditioners, can be incorporated in the ration of ruminants. The spent straw usually has low NV compared with the original straw except the *Agaricus bisporus* harvested washed spent straw which has 5.5–6.0 percent digestible CP and 30–55 percent total digestible nutrients. Spent straw can serve as a maintenance ration for an adult ruminant if supplemented with 200 g maize grains daily. *Pleurotus florida* harvested spent straw could be incorporated in the complete feed of kids without any adverse effect on palatability or nutrient utilisation (Kaur, Wadhwa and Bakshi, 2010).

Limitation and Strengths

The individual-microorganism inoculation experiments were successful in the laboratory under strictly controlled conditions, but failed badly on scaling-up. Industrial treatments which are energy dependent and thus costly should not be recommended for small farmers.

Rumen by-pass Protein Technology for Dairy Animals

In recent years, several technologies have been developed through intensive animal-nutrition research, one of them being by-pass protein feed technology. By-pass protein refers to dietary protein that escapes rumen degradation. The technology aims at decreasing the wasteful production of ammonia in rumen from highly degradable protein meals, thereby increasing the availability of essential amino acids at the intestinal level. The main aim is to increase the efficiency of protein utilization in ruminants for enhanced milk production. Dairy nutritionists try to enhance nitrogen utilization through dietary manipulation to optimise milk production. Manipulation of protein degradation in the rumen is the most effective strategy to reduce nitrogen losses in dairy animals. Losses of nitrogen can be reduced by balancing the ration to achieve optimum ratio of rumen degradable protein (RDP) to rumen undegradable protein (UDP) and increase nitrogen use by ruminal microorganisms. Optimising RDP:UDP also optimises post-ruminal amino acid supply for productive purposes.

Experiences of Applying By-Pass Protein Technology

Methods evaluated for enhancing by-pass protein value. Protein meals are normally fed 'as such' to ruminants in India. However the meals have varying degrees of naturally rumen-protected proteins. The fact that solubility of proteins change, when subjected to special treatments, can be exploited to protect good-quality proteins from rumen degradation. A number of treatments to protect proteins have been tried, e.g. alkali, xylose, heat and formaldehyde. Among these formaldehyde treatment has the advantage of being the most cost effective technology for protecting highly-degradable proteins in the rumen and not adversely affecting animal health and milk quality. This method has been extensively used because of the following advantages:

- Desired level of protein protection can be achieved - under- and over-protection of proteins can be avoided;
- The bio-availability of the essential amino acids can be maximized;
- Does not increase contents of acid detergent insoluble nitrogen (ADIN) and neutral detergent insoluble nitrogen (NDIN);
- Less expensive than heat treatment; and
- Helps to control salmonella and reduce mould growth in feedstuffs.

Patented Technology by The Dairyboard

By-pass protein technology is a new-generation technology, wherein by-pass protein feed is produced by a special chemical treatment, developed and patented by the Dairyboard of India. Regionally-available protein meals are treated appropriately, so as to reduce degra-dability of the proteins in the rumen from 60-70 percent to 25-30 percent, in a specially designed airtight plant. By-pass protein feed supplements have been developed by screening protein meals for their amino acid composition and then developing suitable chemical treatment procedures.

Protein meal identified for treatment is first ground, treated chemically at an appropriate level and then stored for 9 days under airtight conditions. After 9 days of incubation period, protein meal is ready for feeding to ruminants and can be stored for more than a year, without any deterioration in the quality.

Applications

Development of test method. The test method used by the feed industry in India to measure by-pass content of protein meals was based on nitrogen solubility in buffer solutions. However phosphate-buffer solubility values do not compare favourably with *in vitro* or the in sacco protein degradability values. The Dairyboard, in technical collaboration with the Commonwealth Scientific and Industrial Research Organization (CSIRO), Australia and the Australian Centre for International Agricultural Research (ACIAR), has standardised an *in vitro* ammonia release method which gives more accurate values for protein by-passability. In this method, a known quantity of protein meal is incubated under anaerobic conditions at 38 °C for 24 hours in strained rumen liquor (SRL). Protein degradation is measured by analysing ammonia nitrogen level in SRL at the end of the incubation period.

Commercialisation of Technology

A commercial by-pass protein production plant was set up in Baroda Milk Union, in Western India, for producing 50 tonnes of by-pass protein feed per day. The demand for the by-pass protein feed produced in this plant has been steadily increasing. Subsequently, the Dairyboard has set up 12 by-pass protein manufacturing plants mainly under the dairy cooperatives network and has thus commercialised the technology.

Operational Health and Safety Measure

Chemical treatment of protein meal(s) is carried out in a specially-designed airtight plant, so that there is no risk to workers operating the plant. Moreover, workers are advised to wear gloves, masks and safety glasses. The US Food and Drug Administration (FDA) has approved the use of formaldehyde as a feed additive to protect proteins from ruminal degradation, to preserve silages, to maintain animal feeds and feed ingredients free from salmonella and to control fungi. Formaldehyde used in treatment of protein meals is at very low levels and poses no health risk to animals and consumers.

Lessons Learnt

Due to the several advantages of by-pass protein technology, including the cost effectiveness of treated meals and increase in milk yield, the technology has been successfully adopted by the feed industry in India, with NDDB taking the lead in propagating the technology. The Dairyboard is also exploring the possibility for setting up by-pass protein plants in the premises of oil solvent extraction units. The animal nutrition group of Dairyboard is also assessing the amino acid profile of different protein meals available in India, so as to prepare a suitable blend of treated protein meals for improving milk production. The technology could also be employed in other parts of the developing world.

Area-specific Mineral Mixture for Sheep and Goats

Minerals play an important role in improving health, reproduction and production of animals. Mineral deficiency in animals is an area-specific problem and is related to water, soil, feed, fodder and topography. Low productivity regarding reproduction, body weight gain and milk yield are associated with mineral deficiency in sheep and goats on degraded pastures. Deficiencies of Ca, P, Zn and Cu are commonly noticed in sheep and goats in dry zones of India. One of the main reasons for mineral deficiency in sheep and goats is low mineral content of native grasses which constitute the major portion of their diet. Mineral mapping of soil, water, feeds and fodder helps to identify specific deficiency and assists formulating and supplying deficient minerals through production and use of area-specific mineral mixtures. To overcome mineral deficiency in sheep and goats during pregnancy and lactation, supplementation in the form of a concentrate feed and/or mineral mixture, in addition to grazing, may be adopted.

Area-Specific Mineral Mixture

At present, commercial mineral mixtures are prepared and marketed without considering the actual deficiency or excess of minerals in animals of the region. An excess of minerals is taxing to the animal system because of the stress on organs and the extra energy animals spend in their excretion. Also the use of excess minerals adds to the cost of feed. On the other hand, supplementation of minerals deficient in the diet assists efficient utilization of absorbed nutrients, resulting in improved growth, milk production and reproductive efficiency. Mineral mapping of soil, water, feeds and fodders, and animal serum was done to identify specific deficiencies and formulate area-specific mineral mixture. Deficiencies of Ca, P, Cu, Zn and Co were recorded in sheep flocks maintained on degraded pastures of the semi-arid region. The area-specific mineral mixture in powder form, incorporating minerals in the required concentrations, was prepared using calcium carbonate, di-calcium phosphate, cobalt sulphate, zinc sulphate, copper sulphate and common salt. The mineral mixture contained 35.31 g Ca, 13.64 g P, 318 mg Cu, 144 mg Zn, 76 mg Co and 39.0 mg Na.

Mineral Mixture in Pellet Versus Powder Form

Initially, the mineral mixture was developed in powder form and tried in sheep flocks. However, the mineral mixture powder was difficult to feed because the sheep flocks were maintained on pastures without any grain supplementation. Accordingly the mineral mixture was changed from powder to pellet form for easy delivery. Mineral mixture pellets of 5 g, incorporating minerals in the required concentration, were prepared using molasses (1 percent) as a binder. Pellets were dried at room temperature to achieve the desired hardness so that they could be transported and used without breakage losses. The average size of pellets was 2.5 cm in length and 6 mm in diameter.

Salient Findings

The area-specific mineral mixture was tested and demonstrated in six villages involving 12 flocks of 50–60 sheep each. The daily supplementation of pelleted mineral mixture at the rate of 5 g resulted in significant beneficial effects.

Due to deficiency of minerals, wool shedding in sheep is a common problem when grazed on degraded pastures. Sheep suffer from skin keratinisation with lesions around the eyes, legs and head. They are also debilitated and weak and have poor intake. Area-specific mineral mixture can be supplemented to overcome skin keratinisation and wool shedding.

At organised farms, ad libitum concentrate feeding of lambs from 3 month of age causes rickets due to the high consumption of concentrate and low intake of roughage, resulting in imbalances of minerals, specially Ca:P ratio. Such cases can be reduced by restricting the concentrate to 40 percent of the total feed intake and supplementing with an area-specific mineral mixture.

High-concentrate feeding also increases the incidence of urinary calculi due to mineral imbalances, and area-specific mineral mixture at the rate of 2 percent in the concentrates can reduce such cases.

Inorganic mineral mixtures are poorly absorbed and retained in the animal system. Chelated minerals (Cu- and Zn-methionine) in the diet increase mineral absorption and retention and improve growth, reproduction and health. Organic forms of Cu and Zn are commercially available and may be used for formulating area-specific mineral mixtures for improving yield and quality of wool (through increasing the concentrations of Cu and Zn) including, improving crimp, tensile strength and elastic properties. Use of Cu and Zn-methionine in farmer flocks resulted in higher Cu and Zn concentration in wool (4.30 and 172.9 ppm) as compared with inorganic salts (3.50 and 167.2 ppm) and an additional wool yield of 60–80 g.

Adoption of the Technology by Farmers

Sheep and goats in semi-arid and arid regions of the country are maintained on poor quality pastures. Low mineral contents of pasture grasses in dry zones cause mineral deficiencies, ranging from acute to mild, which result in loss of production and reproduction. Supplementation with area-specific mineral mixtures can meet animal requirement for minerals and improve the overall productivity of sheep and goats.

The mineral mixture was made available to farmers of five villages in four districts of Rajasthan. After continuous persuasion the technology was adopted by 60 percent of the farmers leading to a four-fold increase in demand for area-specific mineral mixture. The use of this mineral mixture reduced the cases of reproductive disorders, e.g. anoestrus by 40–44 percent, decreased skin keratinisation by 55 to 60 percent and improved wool and milk yield by 8 to 10 percent and 10 to12 percent respectively in farmers' flocks. In addition, cases of bone chewing in sheep grazing on the degraded pastures also decreased after supplementation.

Some constraints to popularising area-specific mineral mixture in the region are inadequate extension networks for disseminating the technology to poor farmers. Also the farmers have inadequate market access to the mineral mixture. As the beneficial effects of supplementing diets with the mineral mixture are not immediate, this may also discourage farmers adopting the technology.

Initially the technology could be disseminated through Government support and once adopted by farmers, support should be gradually withdrawn. The cost of the mineral mixture is very low (0.12–0.13 Rupee per sheep daily or 3.75 Rupees per month; 1 US$ = *ca* 45 Rupees) and therefore can easily be adopted, even by poor farmers. Many farmers are ready to adopt the technology provided they have easy access to the mineral mixture. The mineral mixture can be made available to farmers, at their doorsteps, through dairy cooperative societies in villages. State extension departments, NGOs, State agricultural universities, Indian Council of Agricultural Research institutions and agriculture extension centres should also be involved in dissemination of the technology.

References

Bakshi, M.P.S. & Wadhwa, M. (2011). Nutritional status of dairy animals in different regions of Punjab State in India. *Indian J. Anim. Sci.*, 81: 60–67.

BIS. (2002|). *Mineral mixture for supplementing cattle feeds – specification.* Bureau of Indian Standards, IS 1664:2002, New Delhi.

Church, D.E. (1984). *Livestock Feeds and Feeding.* Corvallis: O&B Books.

Flegel, T.W. (1988). An overview of biological treatment of crop residues. *In* Kiran Singh and J. B. Schiere, eds. *Fibrous crop residues as animal feed*, pp. 41–45, New Delhi, ICAR.

Garcia, L.O. & Restrepo, J.I.R. (1995). *Multinutrients block handbook.* FAO Better Farming Series No. 45, Rome, FAO

Walli, T.K. (2009). Crop residue based densified feed block technology for improving ruminant productivity. *In* Compendium T.K. Walli ed., *Satellite Symposium on Fodder Block Technology*, pp. 67–73. New Delhi, ILDEX India.

4

Animal Breeding and Reproduction

The domestication of livestock species some ten thousand years ago was a vital step in the development of human civilisation. Over the centuries, domestication evolved into breeding and the genetic improvement of livestock. Nowadays, breeders measure many different animal traits and choose the best animals to be the parents of the next generation. This leads to improvement generation after generation through the increased frequency of desired gene variants in the population. Because the breeding response is cumulative, permanent, and can be spread throughout the production chain, farm animal selection has a great impact on farm animal production.

Europe has always played an important role in improving the world's major livestock and aquaculture species. European breeds are used across the world, and European farmers and breeding organisations are major players on the global market. The European farm animal breeding sector will thus have a great influence on, and therefore responsibility for, the future genetic makeup and characteristics of farm animal populations worldwide.

Growing Demand for Food from Animals

There are some 6 billion people in the world today. Despite declining population growth rates, the world population is increasing by about 80 million a year – equivalent to the population of Germany. Around 95% of this increase is taking place in the developing world. The UN predicts that the world population will reach 9 billion by 2050.

Table 1: Average annual economic gain from farm animal breeding (derived from improved production) in the EU/in Europe

	Europe (Mio €)
Dairy cattle	430
Beef cattle	70
Pigs (Europe)	520
Broilers (Europe)	610
Layers (Europe)	125
Salmon, rainbow trout, seabass/seabream, turbot	80
Total	**1830**

Also growing is the per capita demand for animal food, which should continue to rise for at least 20 years. This so-called livestock revolution is a *demand-driven* evolution. The supply of cereals for human consumption should soon be sufficient to satisfy the demand in developing countries, but the supply of animal-derived foods is far from the mark. The 23% of people living in developed countries presently consume 3-4 times more meat and fish and 5-6 times more milk per capita than people in developing countries. As these poorer people get richer, one of the first things they want to buy is more nutritious and satisfying food, and this generally means more animal protein. Animal product consumption is thus increasing massively in developing countries, and will continue to do so over the next 15-20 years.

Developing countries are also playing a growing role in animal production. Southeast Asia, in recent decades, has tremendously stepped up its pork, poultry, egg, and aquaculture production. Other noteworthy examples are Brazil, for beef cattle, chicken, and animal feed, and Mexico and Argentina for beef cattle and Chile for salmon. Breeding activities are also on the rise in these countries.

Both Asia and Europe are densely populated and have comparatively little land available for agriculture. In contrast, the Americas (and Oceania) have relatively large amounts of arable land and pasture as compared to population density. This creates for Europe and Asia a permanent risk of food dependence on the Americas. It also creates a 'natural' push to develop foods and food products for –and sell them to- the more densely populated areas. Related policies include farming subsidies in the USA, stimulation of genomics and biotechnology research, and a favourable business and social climate for developing and implementing new technologies.

Importance of Animal Production

Mankind has been adapting animals for the production of both food (meat, milk, eggs...) and non-food products (wool, leather, bones...) since domestication started. For some farmed fish and shellfish, domestication is still underway. Animal husbandry for other purposes is expected to become more important. Animals have cultural value; they are used for sports and as companions, for maintaining rural areas, and for medical applications (as models of human disease, potentially as providers of organs for transplantation).

The value of animal production at farm level in the European Union-25 (EU25) is *€132 billion,* amounting to 40% of the value of agricultural production (2004). The EU25 counts some 6 million cattle (of which 23 million dairy cows), 103 million sheep, 12 million goats, 4.4 million horses, 11 million beehives, 151 million pigs, 670 million laying hens, 7368 million broilers, and 285 million turkeys (2003). In addition, the annual value of aquaculture business is •2.8 billion (EU25, 2003). The total arable production includes 400 million tonnes of animal feed production, and about 1/3 of the agricultural area is permanent grassland (EU15). Farm animals in the EU consume about 450 million tonnes of feed a year, of which 140 million tonnes are produced by the compound feed manufacturers. The turnover of the European feed industry is estimated at •35 billion. With EU enlargement, the number of farms has more than doubled (to 17 million), and the proportion of farmers in the workforce has grown from around 4% (EU15) to nearly 8% (EU25).

The EU imports •66.6 billion worth of agricultural products and exports •55.7 billion, so the balance is negative. The role of animals for leisure and sports (e.g. dogs, horses) is increasing. In the EU15, over 1 million people get their income from horses.

Socio-Economic Aspects of Animal Breeding

Output at Farm Level

The *€1.83-billion* annual gain derived from breeding *in Europe* does not include the export of breeding stock (important in poultry, pigs, and salmon). If this is added, the estimated gain is easily twice as high. The costs associated with this economic gain are relatively small in relation to the improvement achieved, as genetic improvement is cumulative and investments in small populations at the top of the breeding pyramid are multiplied down the pyramid to the base level: animals for commercial production.

A Competitive Global Market

Farm animal breeding and reproduction operate in a competitive global market with very low margins. Changes in market shares can occur quickly, especially in areas where a handful of companies serve 90% of the global market. Small differences in performance, cost price, or knowledge implementation (e.g. exploiting genomics) may bring about such changes. As a result, smaller companies might not survive the competitive pressure in a fast-changing marketplace with evolving technologies. This applies notably to ruminant breeding, which is mainly still organised nationally. Here, cooperation between European players is required.

Maintaining the strong position of breeding in Europe is thus not automatic. It will require considerable efforts from breeders, but also the positive engagement of researchers, governments, funding bodies, legislative bodies, and European society as a whole.

A Knowledge-Intensive Sector

The farm-animal breeding and reproduction sector is knowledge intensive. This means not only that the sector exploits knowledge to provide the world with breeding stock, but also that the knowledge developed is or can be used to meet local, niche, or cultural demands inside or outside Europe. The past success of European breeding owes much to its longstanding close ties with universities and research institutes, fostering the dissemination of knowledge to the farm and individual breeder level. In some cases (e.g., dairy cattle) and some countries this dissemination has been very successful, but it still needs to be improved in order to take other situations and rapid changes in animal breeding and reproduction technology into account. The knowledge exploited includes not only genetics, genomics, physiology, reproduction, statistics, and computing science, but also -importantly - economics, animal husbandry, storage and analysis of large datasets, and efficient infrastructure design. Benefits of co-operation over species have been large. The aim is to achieve higher production levels and better product quality while improving animal health, reproduction, other physiological characteristics, and/or feed efficiency. The reproductive characteristics of the animals used largely determine the spread of genetic progress and the organisation of breeding schemes.

Breeding organisations in Europe spend around *€150 million* yearly on research, development, and implementation (in-house or outsourced to institutions such as universities). This includes money spent on estimating breeding values and executing breeding programmes, but not the cost of steps further in the chain, like multiplication and commercialisation. Actors outside Europe (in the USA, China, and Australasia, for instance) are making huge investments in genomics and other new biotechnologies. Research and business conditions are more favourable in the countries concerned than in Europe, and there tends to be a more positive attitude towards progress made possible by new technologies. Europe must also develop knowledge and understanding of new technologies with no immediate application, so that we are well placed to understand their benefits and risks. In developing these technologies, it is essential to ensure close collaboration between research and government as well as transparent inclusion of stakeholders from industry and society.

Emerging Technologies and Societal Concerns

Farm animal breeding operates in a sensitive area because it touches upon important issues: food, health, animals, selection, genes... This sensitivity is reflected in societal concerns related to food safety, ethics, cultural value, genetic diversity, animal welfare, and 'naturalness'. Animal breeders are actively seeking to address these concerns, both internally amongst specialists and through dialogue with policymakers and society. Here are some of the issues that they are addressing:

Sustainability

Breeders have a role to play in promoting sustainable worldwide animal production (a widely discussed concept which emerged in the last decade and was developed notably in the EU-funded SEFABAR project). To play this role optimally over the next two decades, the breeding sector must be both economically sound and attuned to societal needs and demands. This notably means taking into account aspects such as biodiversity, environmental protection, food quality and safety, animal health and welfare... European breeders have been addressing the ethics of animal breeding, notably by producing educational material on this topic and by developing ethical codes such as a Code of Good Practice for Farm Animal

Breeding Organisations (Code-EFABAR). Centring on sustainability, this Code aims to enable breeding organisations to become more transparent.

Animal Welfare

In twenty years, most livestock production (pigs, poultry, dairy cattle, beef cattle) will probably be in large-scale units, though still mainly family owned. Animal welfare in this situation will be a high priority and animal breeding can contribute to it. In this context, attention should be paid to developing efficient information management systems for health monitoring, health detection, etc. Similarly, very little is known of the welfare of the domestic aquaculture species today.

New Technologies

Globally, there is active competition for new technologies that may directly or indirectly affect the future of animal production, including breeding. Some non-European countries (e.g. New Zealand, the USA, Argentina, Brazil, China) are rapidly developing research on - and in some cases implementation of -new reproduction and cloning technologies (somatic cell nuclear transfer, i.e. "Dolly-type" cloning) and genetically modified animals. Other foreseen applications of the new biotechnologies in animals are in the medical field: animal models, animals as bioreactors, and animals for xenotransplantation. Although the development and possible use of such applications are beyond the direct scope of breeding organisations, Europe must be in a position to objectively evaluate these technologies and consider their potential.

In conclusion, funding of research on new technologies should continue in Europe, even if we cannot predict today whether we will need this expertise to develop our own future (and what that future will look like) or to protect ourselves in global trade negotiations. This funding should be from non-commercial resources (national, local, European), so that Europe can make a real and unbiased choice when the time is right.

Transparency and Dialogue

Transparency is of the essence, both in breeding and in the development of new technologies and pathways. Also essential is a continuous dialogue with the public as technologies are being developed. We need to discuss how increasing knowledge and the exponentially growing power of computer analyses can be applied in order to achieve balanced breeding effectively.

We need to tackle questions such as: can finding the right balance enable us to meet global challenges? Can genomics be a powerful tool for this purpose? How can breeding knowledge and the organisation of breeding be exploited in order to satisfy the growing demand for animal products and to meet specific needs in relation to emerging economies, food safety, food supply, and the maintenance of diversity in small breeds?

Control over Future Decisions

As breeding is a competitive global market, cloned or transgenic (genetically modified) animals developed outside Europe or products derived therefrom are bound to find their way eventually to the European market. Currently, however, knowledge of technologies such as cloning or transgenesis (gene transfer) in farm animals is rapidly disappearing from Europe. For Europe, the only way to stay in control of its own decisions is to keep pace with international developments. Europe may want to close its borders to technologies it does not appreciate, but for this it will need sufficient knowledge and high-level scientists to support its claims internationally, e.g. in World Trade Organisation (WTO) negotiations. There are other good reasons why Europe should develop further its expertise and skills in new technologies. The importance of this type of research for the non-food area is potentially so great that Europe cannot afford to miss out on new developments. Ultimately the efficiency of new technologies may increase until there is no impact on animal health, animal welfare, or the environment.

Links with Animal Populations and Culture

Breeding is working with animal populations. In the case of most species this means that there is a close link with national breeds and that breeding is often organised nationally. Animal breeding is not only about animal products. Especially in the case of extensively farmed grazing animals, it is also about domestic animal biodiversity, food security, rural landscape management, and cultural values. These aspects are important in Europe.

Organisation of Farm Animal Breeding

Farm animal breeding is organised at different levels, from farms to breeding companies to breeding organisations up to the level of regulatory bodies providing national and international (e.g. EU) regulations. The organisation of farm animal breeding differs significantly according to the species. In poultry and fish farming, there are relatively few breeding organisations worldwide, as breeding work is knowledge intensive and relatively expensive. In the case of ruminants, farmers are often organised co-operatively for technical services, representation activities, and breeding. Pig breeding is in an intermediate situation.

In recent decades, only collaborative or specialised organisations have been able to breed livestock effectively, taking into account the latest scientific developments, using powerful computing systems and sophisticated estimation programmes, maintaining huge pedigrees of animals with their performances and biodiversity, and exploiting reproductive technologies. Yet there are no very big players in animal breeding. The organisations involved are (small or) medium enterprises or small units in larger organisations.

Poultry and Aquaculture Species

It is possible to concentrate breeding of small farm animals because the number of offspring per parent is high. This is what has happened. Margins being extremely low for eggs and meat from these species, breeding costs have been reduced as much as possible. In many cases former breeders have become multipliers (a step in the dissemination of breeding stock), outsourcing the selection work to another company. In poultry, for layer and broiler chickens and turkeys, 2-3 organisations are responsible for over 90% of the global breeding stock in each market. Aquaculture, a young sector, is developing in the same direction. Here, the number of species that can be farmed is rapidly increasing. Selective breeding is being applied to salmon, trout, seabass, seabream, and turbot, as well as other aquatic species such as shrimp and oysters. Breeding programmes for recently domesticated cod strains are now beginning. It is necessary to develop breeding and reproduction of additional species in aquaculture, so that closed selective breeding schemes can be initiated and the over fishing of wild stock can be avoided when the demand for aquaculture products increases.

Ruminants and Pigs

In the breeding of ruminants (cattle, goats, sheep) and still, to some extent, pigs, the animals' limited reproductive output means that a large number of breeding units (on farms) are needed in order to disseminate desired characteristics from elite animals at the top of the breeding pyramid. In addition, cows on dairy farms are the mothers of potential future breeding animals and are hence also a part of the breeding chain.

The breeding of small ruminants such as sheep and goats is also specialised. The farmers of these species often live in marginal areas, far from the market and with a more limited access to technology than in the case of other species. Small-ruminant farming is often organised in a weaker co-operative system than for cattle. The importance of these species is likely to increase in the future because of their important environmental impact. New technologies in breeding and reproduction have not yet had much impact and should therefore be stressed.

In pigs, many farms are involved in disseminating top breeding animals to the farmer who grows the pig. Breeding programmes are often based on crossbreeding schemes integrating a dissemination level and a core breeding level.

In ruminants and pigs, most breeding organisations are cooperatively owned by farmers. The organisations based in Europe are world leaders for their species. They are often SMEs operating in their native language, organised in national umbrella organisations. For disseminating desired genotypes, the national or local organisations of ruminant farm owners are internationally co-ordinated by a single international institution, based in Europe, created to provide services, mainly in animal breeding, on a larger, international scale. The largest pig breeding organisations in this sector are private or cooperative European organisations with a major impact worldwide.

Ownership

The purchase price of an animal includes its breeding rights, so the reproduction rights for an animal belong to the animal's owner. There is no animal equivalent of plant breeders' rights. The population structure of breeding makes such protective systems very sensitive to 'cheating'. Private investment in research, on the other hand, should be stimulated by dedicated protection of relevant findings.

It is important to strike a good balance between knowledge sharing and the protection of newly developed knowledge and tools. Patents are not commonly used in breeding – existing methodologies and knowledge should not be subject to 'broad claims' and 'new claims' that would hamper access to methods already applied or described and discussed openly at round tables and conferences, in scientific magazines, etc.

Structural Change

The number of main organisations active in farm animal breeding and reproduction varies considerably in the EU. Yet across the EU there are many thousands of smaller pedigree breeders, most of whom participate in larger breeding programmes organised by some of these main organisations. In the breeding of all species there will likely be a trend towards concentration, i.e. towards fewer, larger breeding enterprises.

Regulatory Aspects

The organisation of animal breeding and reproduction is under permanent supervision by - and interaction with - regulatory bodies providing national and international regulations (e.g. EU authorities and ICAR, the International Committee for Animal Recording). Interaction with these bodies is crucial to promoting the harmonious development of animal breeding and reproduction practices, especially in ruminant farming with its open breeding structure.

A Vision for Sustainable Breeding Reproduction

Animal breeding and reproduction have a great contribution to make to a future sustainable animal agriculture. Many opportunities are open to the animal breeding and reproduction sector for improving the biological and economic efficiency of food production and increasing food supply. These opportunities are particularly attractive in a sustainability perspective. The sector is intimately linked to all three pillars of sustainability: *Consumer, Environment, and Economy*, through its interactions with societal developments, its implications for biodiversity and the environment, and its contribution to economic growth. How these three pillars will develop in the next twenty years and how sustainable breeding in the EU will fit into the picture are questions that must be addressed.

Sustainable breeding and reproduction means balancing:

- safe and healthy food
- robust, adapted, healthy animals
- biodiversity
- social responsibility
- a competitive and distinctive

Europe and must include new prospects for animal production in Europe. When it comes to developing new technologies, transparency about the developments and an open dialogue with society are prerequisites.

Integration into Animal Agriculture

Traditionally, animal agriculture and aquaculture are fragmented into distinct activities and focuses: animal nutrition, animal waste management, food science applied to animal products, animal welfare, animal management, and animal breeding and reproduction. Given the challenges facing animal agriculture and aquaculture today, we believe that a much higher degree of integration is needed. In our vision, integration will affect all levels and will certainly influence the future of animal breeding and reproduction, as outlined below.

Safe and Healthy Food

Product Quality

In a "fork to farm" production system driven by consumer needs, quality in the broadest sense (food safety, nutritional value, sensory and technological quality) becomes one of the most important drivers. Safe and wholesome food is clearly the first requirement. Beyond safety and for all animal products, product quality improvement will be a major issue. This is a significant shift from the food sector's position in the last century, when improving the level of production was the obvious goal of a supply-oriented production system. In the modern context, nutritional value and human health features (e.g. the fatty acid composition of meat and milk, the nutritional quality of eggs) can become targets for animal breeders, as can sensory qualities such as tenderness, flavour (importantly, this includes boar taint), visual appeal, and processing characteristics. The importance of breeding systems and new technologies in protecting the quality and enhancing the economic efficiency of local and typical productions deserves full consideration.

Zoonoses

Farm animal breeding and reproduction have a role to play in decreasing the incidence of zoonoses. The opportunity to breed animals that are resistant to spreading zoonotic diseases or have a generally improved immune system should be seized, as it will greatly improve food safety. Both animal and human health will benefit.

No Residues in Food

Ideally food should contain no residues. This is an important aim. By selecting animals that are more robust and able to adapt easily to the production environment (e.g. feeding system, climate, housing/grazing system), it will be possible to reduce the need for medicines and thus the risk of residues in animal-derived food.

Safe Gene Dissemination

The safe dissemination of animal genes is a major issue. The risk of transmitting diseases through animal, semen, egg, and embryo transport has increased with increasing farm size and the internationalisation of trade. It is essential to minimise this risk. Artificial Insemination and embryo transfer have contributed a great deal towards this goal, but there is room for further reducing the disease transmission risk and ensuring safe transport of genetic material by improving the health guarantees of this material. The role that existing and new reproduction technologies can play in the (international) transport of breeding material should thus be weighed against the ethical dimension of using certain technologies (e.g. embryo technologies). Specific Pathogen Free (SPF) production of breeding animals results in animals free of certain diseases, which in turn leads to improved human and animal welfare.

Robust, Adapted, Healthy Animals

Animal Welfare

Breeding organisations ensure the health and welfare of the animals they keep and select. They are engaged in the search for selectable traits that are indicative of species-specific animal welfare. New biological insights into brain function, the genetics of behaviour, and physiological indicators of stress and well-being will provide new tools enabling breeders to handle welfare traits more objectively than at present.

Adaptability

Breeders produce animals for a wide range of production environments, from extensive and/or organic systems to more intensive systems on larger farms (which continue to get even larger). Breeders want to be able to improve the level and efficiency of production in each of these environments. It is essential that animals remain healthy and productive under this wide range of environments and with – 16 of 30 - February 2006 contribute to defining appropriate levels of biodiversity, identifying in small populations 'rare' gene variants whose preservation should be reduced human interference (labour costs are the main costs in farm animal production, along with feed). It is now common for breeders to test animal performance in more than one environment, but new knowledge on how animals adapt will enable breeders to take novel aspects of production into account more effectively in breeding programmes. Over the next twenty years there will be more large-scale units and new technologies for which an efficient information management system will be developed. Aims will be to facilitate selection of animals for adaptation to these systems and to use technology to support balanced breeding programmes.

Domestication of Fish

Most farm animals (and companion and sporting animals) were domesticated thousands of years ago. This is not the case of aquaculture species. Here, the domestication process has just started. A lot of work has to be done towards improving the reproduction of aquaculture species and towards domesticating species that are now only caught in the wild.

Disease Resistance

Breeding may contribute to robust and healthy animals by selection for more general and specific disease resistance as is done for aquaculture species today. This should result in less use of medicines. Optimising the use of medicines and vaccines according to an animal's genetic makeup represents a significant development opportunity.

Balanced Breeding and Biodiversity

Balance

It is necessary to further increase production levels so as to lower cost price and to ensure an adequate supply of meat, fish, eggs, and milk for a fast-growing human population demanding more of these products. Yet especially in Europe, breeding cannot be driven by production efficiency alone. In defining breeding objectives, it is necessary to strike a good balance between the production level, animal physiology (welfare, behaviour, health, reproduction), and population fitness. By learning more about animal genes and physiology and by exploiting greater computing power, it should be easier to balance these multiple breeding objectives. To achieve this optimally, we need more knowledge and better tools in areas such as the basic biology of traits and their interrelationships and data handling. Also essential is the improvement of genetic and reproductive tools with regard to precision, ease of use, and cost.

Biodiversity

Breeding programmes are designed to make optimal use of existing genetic variation between and within populations. Breeding organisations must contribute to maintaining genetic diversity in their breeding populations. They must monitor and control the rate of inbreeding and genetic drift. In animal populations with an open structure (e.g. cattle), the 'effective' population size is much smaller than the actual population size. A research opportunity is to better understand the interaction of biodiversity and genetic variation. This will foster knowledge-building in diverse areas, such as the co-evolution of hosts and pathogens and animal adaptation to climatic differences or variable nutrient availability. It will also prioritised, and protecting genetic diversity in populations used for large-scale production. In the coming years, more attention will have to be paid to managing (as distinct from conserving) animal genetic resources, particularly given the sometimes narrow genetic base of pigs, poultry, and dairy cattle, and the concentration of breeding strategies and selection in very few hands.

Furthermore, it is important to bear in mind that biodiversity preservation is the key to preserving future breeding opportunities. It is essential to maintain sufficiently diverse gene pools because we cannot foretell the changes that might affect the social and economic driving forces of animal agriculture. Nor can we predict the evolution of biodiversity-related ethno-zootechnical issues that influence the link between farm animals and society. The preservation of biodiversity is also important for conserving and improving local breeds, which are very significant in marginal areas. The need is to protect these breeds by increasing their production and standardising the quality of their products. Objectives will be to defend the environment of marginal areas and to maintain the efficiency of animal farming in less favourable areas.

Social Responsibility

Environment

There is scope for selecting animals better suited for maintaining biodiversity under changing climate conditions. A major achievement of animal breeding in recent years, and one with a highly favourable effect on the environment, is the improvement of feed efficiency. When animals make more efficient use of nitrogen and phosphate, this minimises the output of these elements into the environment. The main contributor to improved feed efficiency is improved feed conversion. Yet there is room for further improvement of this process, especially in farmed fish. Progress will require new knowledge on the genes and pathways that may contribute to improving the European environment while an animal's production level is increasing. Opportunities for feed efficiency improvement exist in all farm animals, from fish to cattle. This requires better knowledge of the digestive system and of nutrient metabolism. Of particular interest are the genes whose expression is regulated by nutrition. When exploiting such knowledge in animal selection, it is necessary to take into account animal product quality and both beneficial and unfavourable environmental outputs. Animal breeding contributes to the environment in yet another way: through selection of animals well adapted to their environment and interacting favourably with it by grazing and pasturing. Such animals contribute significantly to ecosystem maintenance. The selection of genotypes for the maintenance of the environment is fundamental in many European geographical areas, especially mountain and Mediterranean areas.

Animal Integrity

Animal integrity is defined as "the wholeness and completeness of the species-specific balance of the creature, as well as the animal's capacity to maintain itself independently in an environment suitable to the species." Most farm animals have been domesticated for thousands of years and as such have been adapted to an environment of co-existence with humans. They are easier to handle and able to reproduce in captivity, under conditions where food is abundant and there are no direct predators. Breeders must keep a balance between the intrinsic characteristics of domesticated species, their welfare, and improved production.

World Development

The global demand for animal-derived foods is increasing. European breeding organisations play a major role in providing breeding stock globally. For pigs, aquaculture species and poultry, which are reared in relatively controlled environments, this has generally been a huge success. The history of cattle breeding, however, is littered with mistakes. High-productivity European breeds are not always suited to more extensive environments characterised by a different and sporadic food supply and by various climatic and disease challenges. It is possible to learn from these mistakes. In global development programmes, the knowledge and organisational skills of European breeders could be adapted to meet local needs in developing countries. New breeding schemes require multiple-character selection based on both production traits and other important traits such as resistance to disease (sanitary quality and animal welfare), fertility, and longevity (animal welfare). At the same time, there is a need to conserve biodiversity through the development of appropriate breeding strategies in rare ruminant breeds.

A Strong Breeding Sector

A prerequisite to breeding in Europe is the survival of the breeding sector. Major companies have premises for breeding and reproduction outside Europe, where research can often be more easily funded and carried out.

There is a constant push to move away from the head offices in Europe in order not to lose market potential and ultimately lose out in the competition. Every year breeding companies disappear or amalgamate. When companies move away, they will not come back. An important challenge and opportunity is to keep the breeding climate in Europe challenging and interesting enough to maintain businesses here. Breeding needs profitability at all levels and requires active participation of farmers in order to test animals for desired characteristics and to disseminate those characteristics. This is feasible only if such endeavours are economically sustainable. It is crucial to preserve the livestock system by maintaining and/or increasing the efficiency of breeding also in small-scale animal farming. Many of the new EU member countries have a tradition of small-scale aquaculture, which needs to become more efficient in order to compete. Breeding is a global activity, and European businesses have gained leadership in the breeding of several farm species. Yet the market remains competitive and as technology developments continue at a breathtaking pace, European breeders must not be complacent. There are structural weaknesses in Europe compared to other global players, as breeding is often organised nationally. Given the existence of 25 different countries in the current European Union, this is a natural disadvantage for European animal breeding. Further strategic partnerships in research, development, and business across Europe will be increasingly important. Research opportunities will arise in operational genetics, numerical biology, animal recording, data management, and high-throughput biological research. If Europe is to retain its competitive strength, there must be sufficient support from research funders to ensure that the basic and strategic research opportunities are seized. European public funding should play an important role in this context, helping to foster

integration of research activities and co-ordination of research and innovation policies.

It is essential to bear in mind that the changing world will demand careful but *fast* adaptation of farm animal production systems and hence of farm animal research and breeding in its broadest context.

Transparent Public Engagement

New technologies, goals, or practices will be part of the road to the future. In some parts of the world people are already applying food production technologies that Europe considers unacceptable. This trend towards parallel development paths may well continue, so Europe must develop alternative approaches if its agriculture is to avoid facing even greater competition than it does today from cheaper imported foods. One promising pathway might be a trend towards differentiated foods (e.g., speciality foods, organic foods), as food production in Europe is likely to develop towards quality instead of quantity. Alongside the development of new technical solutions to breeding problems, it is important to involve specialists in ethics to support careful interaction with society as a whole. It is vital to realise and take into account which breeding activities are in line with what European citizens want animals to do and produce and what they consider inappropriate. Transparency about breeding work and research developments is a key issue for farm animal breeding scientists and breeding organisations. Animal breeding strategy over the next twenty years must include a commitment to communicate with civil society. In the past, a lack of information has sometimes given new breeding technologies a negative public image.

A Flexible, Open Environment

For European breeding organisations, competition at the global level is strong. Unlike breeders in the largest competitor country, the USA, European breeders must deal with a multiplicity of scientific, funding, and regulatory environments within Europe. European breeders need a flexible, open environment in order to continue their business and to develop and implement the schemes needed for the future. They are therefore in favour of initiatives such as regular and timely consultations with regulatory agencies, dedicated to finding workable yet safe solutions for animal transport and new technology implementation (guided by ethical and safety considerations), investigating specific conditions for minor species or species under development, and creating the same conditions for imported material as for European material.

Well-Educated People

For a knowledge-intensive sector, the education of students and hence of future science and industry leaders is important. The growing complexity of animal breeding makes this especially true. Today the sector requires a knowledge base combining largely quantitative science with molecular genetics and genomics, as it pogresses towards the day when it will be an industrial user of predictive biology. Education, in the future, should be a more collaborative endeavour where public and private bodies cooperate more closely.

Rural Areas

Farm animals, especially ruminants but also horses, have an important role in maintaining the European rural landscape, e.g. in keeping areas clear to prevent fires. Only a few local breeds are now still part of that landscape and add to its diversity across Europe. Therefore, we need to maintain those breeds that fit with the landscape and, in harmony with the environment, yield high-quality local products. We need to breed them to maintain the landscape whilst improving their economic sustainability. To reach these objectives, it will be necessary to consider breeding goals and technologies for extensive systems. The value of the resulting breeding system will be enhanced by the enormous importance of extensive livestock farming not only for production but also for its environmental and social benefits. Aquaculture production has another value for the rural areas as it is often situated in such areas and can therefore assure that work-places are secured there.

Regional Products

Europe is a diverse continent with typical products related to local breeds, local environment or climate, or local habits or history. Consumers often go for a value-for-money product, but increasingly they appreciate the perceived added value of regional products. In many situations in Southern Europe and mountain areas, breeding will have to be planned to improve the distinctive qualities and local characteristics of animals. This will lead to a differentiation of food products.

Social/Cultural Aspects

European animal production is also important for the vitality of the countryside. When farmers disappear and holiday homes take over, the social infrastructure can suffer (e.g. schools, shops, public transport). Cultural values may open a market for some higher-value niche animal products, appreciated by consumers when the origin of the food is part of its attraction (e.g. when eating for pleasure in a restaurant or cooking a special meal).

Diversity of Benefits

Animals for Many Purposes

With the standard of living increasing and more time available for most people, animals for pleasure and leisure are a growing industry. In addition to companion and sporting animals, there is a growing trend towards 'hobby' farming. Animals will also be bred for social purposes, e.g. for 'animal therapy' or for helping to rescue people in an emergency. Breeding for the right type/character of animal for the right purpose may be a new, promising prospect for breeders.

Fibre and Skin

Wool, leather, and skin are among the traditional farm animal products. They used to be necessary items for clothing, but nowadays they are used for their 'naturalness' (e.g. wool) or for the specific advantages of the product (e.g. leather). Breeding sheep or other fibre animals for typical products, like natural coloured wool, is an increasing niche market.

Animal Biotechnology

The combined advances in genetics, embryology, and stem cell research open the prospect that high-added-value products for agricultural, medical, and technical applications will soon become a reality. The major applications will be in the use of highly specialised animals for drug production (e.g. in milk), for xenotransplantation, or as animal models in research on human diseases.

In the future, animals may become precious tools for modelling human disease and developing new therapeutic strategies. Such approaches will both complement and inform studies based on human cell/tissue cultures. To make such applications possible, it will be necessary to develop robust embryonic stem cell lines capable of self-renewal in culture and to refine the technologies used for cloning and transgenesis. These same technologies will be valuable research tools enabling scientists to gain deeper understanding of livestock and companion-animal embryology and reproduction. A challenge in the next two decades will be to integrate and potentially exploit these novel technologies in a society-friendly manner. For this it is essential to build the knowledge base and expertise required to provide technological guidance, quality control, and safety.

An Agenda for Research

Quantitative Genetics and Operational Genetics

Some inherited traits, such as milk yield or daily gain, are controlled by many genes. In an animal population, different animals will have different combinations of gene variants affecting a given trait. This leads to quantitative variation of the trait within the population. Such traits and the underlying genes are the focus of quantitative genetics. Over the past fifty years, most improvement in the quality and efficiency of farm animal production has been due to the application of quantitative genetic methods. For instance, the International Bull Evaluation Service (Interbull) maintains and provides access to a huge body of genetic animal evaluations from the most advanced countries. Its ranking of bulls according to their 'genetic merit' facilitates the genetic market worldwide.

Operational genetics is the application of genetics to breeding, for example in designing breeding programmes and optimising selection schemes to ensure maximum genetic gain while minimising inbreeding or undesirable trade-offs of selection.

If breeders are to maintain and build upon the enormous genetic progress made in past decades, they need the support of research focusing on the further development of complex statistical methods and computer algorithms for coping with multiple traits. They need information from new sources such as genomics and the study of how animals adapt to specific environments. Attention must also focus on the more complex, non-linear, and difficult-to-measure traits such as disease resistance, longevity, and robustness, as well as on new and emerging traits. Alternative selection schemes should also be envisaged for the analysis of non-additive genetic variation.

Advanced Modelling for Management Purposes

Quantitative genetics has a favourable 'side effect' of which people are not always aware: extensive modelling of complex phenotypes (biological traits) provides interesting and important information about environmental influences on these phenotypes and makes it possible to predict unobserved phenotypes. It will be important to incorporate this and all relevant knowledge into advanced management tools, which will notably increase the return on investment of extensive data recording schemes. Current efforts in the field of milk recording (test-day models) are just the start of a whole new range of research opportunities. Future research in statistics and computing sciences will be required to optimise current methods and procedures. As an on-farm management system develops, a challenge will be to link it to more integrated, multi-level systems.

Phenomics

Phenomics means the measurement of animal phenotypes. Currently animal breeding programmes can only improve measurable traits or traits genetically related to measurable traits. They rely on characterising animals for production, quality, and welfare traits on the basis of limited numbers of laboriously collected records. The entire selection system for ruminants, through which many traits have been improved, is based on this traditional approach, embodied in a manual used throughout Europe and on other continents: the 'ICAR Guidelines for Animal Recording'. In the future such methods are likely to be complemented by high-throughput automated recording systems, on farms or in processing factories for example.

An important future research focus will be the estimation of carcass meat content by means of automated image collection. Accurate measurement of many other meat traits (e.g. colour, fillet yield) currently requires animal slaughter, but new techniques have emerged (Near Infrared Spectroscopy (NIR), ultrasound and computer tomography) that will allow such traits to be recorded on the breeding animal. This will increase the accuracy of selection and hence, the genetic gain. Ultrasound techniques were successfully implemented in pig and beef cattle breeding in the 1970s and can be further developed in other species. In dairy animals, more sophisticated techniques for the recording of milk yield and composition are necessary in order to continue improving the production efficiency and nutritional quality of milk and milk products through selection. Much remains to be learned about how biological attributes and genetic features influence many sensory traits of meat. For example, are variations in meat texture due to muscle fibre number, fat content, the amount of connective tissue, or to all of these factors? Before breeding programmes can tackle such traits, we need to know which measurements are needed to describe them and how to interpret those measurements.

Identification and Traceability Technologies (ID)

Animal identification is the cornerstone of selective breeding programmes. In the absence of accurate identification, no pedigree can be established and all information on the animal is lost.

The ability to trace an animal from birth through to a finished product (meat, milk, eggs, wool, leather) is crucial in the battle for safe food, in controlling contagious diseases, identify fish that have escaped into the wild, and in the effort to inform consumers about food attributes such as country of origin, animal welfare, and genetic composition. The development of animal breeding systems for ruminants and pigs will make full use of technical improvements in physical identification: Ear-Tag systems, Radio-Frequency Identification (RFID), electronic boluses... Long-term research should be devoted to all these methods. Another approach that has huge potential is the use of genetic markers for individual identification. This approach should prove particularly useful in aquaculture and poultry farming, where physical identification at birth is difficult to impossible, but it is also applicable to other species. Research is thus needed in areas such as high-throughput genotyping (i.e. genetic marker typing) and DNA chip technology in order to address issues of practicability, cost, and access. It is also urgent to explore how new technologies might be applied to traditional identification systems having an immediate and wide application. Furthermore, all actors in the field must collaborate towards ironing out the relevant legal and regulatory aspects.

Genomics and Beyond

An organism's genome is its genetic material (DNA). Animal genomics, broadly defined, is the study of animal DNA, its organisation into genes, the roles and interactions of gene products (mostly proteins), the control of gene expression, and ultimately all downstream impacts on animals, their traits, and their interactions with the environment.

The sequencing of the human genome was a milestone in the development of modern biotechnology, opening new avenues for understanding the control of complex traits at a molecular level. The genomes of several farm animal species (cattle, pig, rabbit, and aquacultural species) are also being sequenced and analysed. This will facilitate the integrated analysis of biological functions.

At sequence level, genomic research is yielding genetic markers for selection, pedigree control, and traceability. It is facilitating the identification of trait genes for use in breeding programmes. Beyond this level opens the whole new realm of "omics" research: transcriptomics (focus: expressed genes), proteomics (proteins and their functions and interactions), and metabolomics (metabolites and metabolic pathways). This research is bridging (part of) the gap between 'structural' genomic knowledge (markers, maps, sequences, genes) and everything we hope to learn by exploiting it. It provides a bridge to nutrition, health, and more. It is yielding precious knowledge on a wide range of processes related to health and food safety, such as acquired resistance to contamination through gut health or the long-term effects of an altered maternal diet (foetal programming) in both animals and humans. The application of such knowledge (in marker-assisted breeding programmes, for instance) will contribute to improving animal health and food safety, animal welfare, and the biodiversity of breeding populations. In the long run it will boost the competitiveness of European farm animal breeding.

Functional Genomics

This area includes transcriptomics and proteomics. It aims to elucidate gene functions. Transcriptomics uses powerful tools (e.g. microarrays: arrays of material resulting from gene transcription, the first step in gene expression) to study simultaneously the expression of many (ultimately all) genes in a genome in various situations (specific developmental stages, environmental stress…). Proteomics exploits various technologies to study which proteins are present in -or secreted by - different cells under different conditions, and also looks at protein interactions.

Functional genomics is used to study genes that contribute jointly to a specific function as part of the 'interactome' (the whole set of molecular interactions in cells). This notably includes networks of interacting proteins, structural and mitochondrial DNA, and inhibitor RNAs (RNA is quite similar chemically to DNA; it plays multiple roles, notably in gene expression and its regulation). By combining multiple approaches and tools, it should eventually be possible to build models linking together large numbers of genes and their products, in order to understand the impact of gene variation on animal traits, mediated through the relevant metabolic pathways. To achieve this, a considerable research effort is needed, yielding better understanding of animal biology on the basis of knowledge obtained from the genomes of model organisms and humans.

Exploring Within-Species Variability

To allow effective selection of animals via genomic approaches, it is necessary to explore within-species variability and to quantify factors such as linkage disequilibrium (the tendency of genetic features that are close to each other on the genome to be inherited together, and thus frequently to be found together in the animals composing a population). Effective genomic selection requires the development of tools for genotyping animals by marker typing or direct sequencing. It requires improved tools for phenotyping animals more exhaustively and quickly so as to identify genes that control variation of multiple traits. We also need to develop strategies for predicting phenotypes on the basis of genotypes. Here research should notably focus on the genetic control of variation in gene expression and the identification of genome regions undergoing epigenetic modifications such as imprinting. Given the strong negative impact of inbreeding on reproductive and health traits, we need to find ways to reduce the inbreeding coefficient in pigs, poultry, aquaculture species, and dairy cattle.

Exploring Between-Species Variation

By learning more about between-species variation, it will become clear which vital processes program which functions in animals. This will enable breeders to identify animals yielding healthy animal products without this being detrimental to their own health. With genome-scale sequencing of domestic species, between-species comparisons (including comparisons with mouse and human sequences) will make it possible to identify variability 'hot spots' in genes and/or genome regions so as to focus the search for single nucleotide polymorphisms (SNPs) responsible for substitutions in protein sequences in domestic animals (SNPs are sites where a single 'letter' in the 'text' of a gene is variable).

Numerical Biology

Future biology will increasingly require the analysis of large volumes of data (e.g. genomic data). Better numerical methods for analysis and modelling are needed to address a whole range of biological problems from the molecular to the ecosystem level. Identifying functional variants in DNA sequences or differentially expressed genes on microarrays, analysing host-pathogen interactions in farm animals populations, are just a few examples among many. There will be an increasing need for better numerical models of biological (sub-)systems. For example, scientists currently use separate models to predict protein folding from amino acid sequences and the spread of disease from epidemiological data. A more elaborate model would predict the impact of designed changes in an immune response gene on the frequency of disease epidemics in domestic animals. As the need for numerical analysis and modelling spans all areas of biology, the field of animal breeding and genetics will have to adapt to its own needs some new methods developed for other purposes, in addition to contributing approaches specifically designed for livestock problems.

Biology of Systems and Traits

Currently the basic biology (genetics, biochemistry, and physiology) underlying the variability of most traits of interest to animal breeders is poorly understood. This is an obstacle to controlling physiological processes that influence such traits. Molecular genetics will yield abundant novel information on the fundamental regulation of such processes. Here research must aim to provide detailed operational understanding of the pathways involved. Many gene products influence several metabolic pathways, and most metabolic pathways are influenced by several gene products. This makes for very complicated physiological network patterns. In addition, whole sets of genes can be switched on or off via so-called 'epigenetic' mechanisms (DNA methylation, histone acetylation, RNA interference) participating in complex regulatory networks. Such mechanisms are increasingly found to play significant roles in phenotype development and transmission. Control of a single gene may thus cause unwanted side effects when the physiological network around it is not sufficiently understood. Future research must focus on dissecting the genetic basis of traits of relevance to sustainable animal breeding and the interrelationships of these traits.

Whole-Animal Biology and Gene-by-Environment Interactions (GxE)

Understanding individual traits is not enough. What matters is the biology of the whole animal and how an animal's genes interact with its living environment to determine its quality of life and performance. Biological research is already moving towards predictive biology. A first step is to model the various systems that constitute the biology of a cell (creating an e-cell) so as to predict cell responses to environmental change (e.g. altered nutrient supply).

The next step is to build models that describe cell-cell interactions in tissues and finally to build upward from tissues to a whole animal. This is a considerable challenge, but it is undeniably the direction in which biology is heading. High-throughput laboratory methods now make it possible to measure many components of a system in parallel, and these approaches complement the traditional whole-animal research supporting farm animal production. With modern research techniques, scientists will be able to revisit 100 years of animal science with a view to expanding knowledge of all biological mechanisms. Future research must focus on developing an integrated understanding of the lifetime functioning of whole animals and their interactions with the environment. At each step it will be essential to evaluate progress in the light of breeding reality. What breeders need especially is a detailed understanding of relationships between performance, health, welfare, and environmental impact combined with cost-effective, robust, reliable ways to measure traits reflecting these relationships in animal populations.

Population Biology

Future qualitative and quantitative genetic technology will give breeders much control over the genetic makeup of individual animals selected for breeding. In other words, what can be selected with increasing precision is the genome of an individual. Yet population-level effects also deserve consideration for at least three reasons. Firstly, social interactions in groups of animals are associated with behavioural repertoires that are important for animal welfare and for proper functioning of the group: maternal behaviour, dominance and aggression... To the extent that such interactions are linked to production traits, selection could have negative side effects when the physiology of social behaviour is not sufficiently understood.

Secondly, the production potential inherent in an animal's genes is variably expressed according to the nutritional, climatic, infectious, and social environment. This is called environmental sensitivity or phenotypic plasticity. It provides a measure of genotype-environment interactions at the level of an individual animal. Future genetic technology will offer increasing control over environmental sensitivity, and again side effects are likely if interaction mechanisms are not properly understood at population level. It is important to take into account the dynamics of the environment in which future breeding animals will have to produce. The environment will change in the next twenty years and strategies for the genetic improvement of animal populations must adapt to these changes. Also important is the development of areas such as the improved management of pastures and silvo-pastoral systems for increased overall efficiency in animal production. Research should focus on defining, through population biology, which will be the most efficient genotypes for predictable environments.

Thirdly, disease transmission is perhaps the most crucial population-level issue. Infectious disease involves two (or more) 'actors': host (the farm animal) and pathogen. These two players interact through infection (pathogen to host) and immunity (host to pathogen), and the pathways involved in both processes are under genetic control. Future genetic technology will provide much knowledge about regulation on each side of such a system, but the connection between the two, i.e. the host-pathogen interaction, is likely to be the crucial element in getting the system under control. Exploring this connection means focusing on epidemiology in relation to pathogen virulence and host resistance. It requires studying the co-evolution of hosts and pathogens. Research exploiting emerging technologies is bound to provide much knowledge on the regulation of the immune system at the individual animal level, but turning this knowledge into operational control will require linking it properly to population-level epidemiology. Future research must focus on understanding how variation in a population affects the overall performance of groups of animals and how changes in biodiversity are likely to influence the control of infectious disease. This will ultimately lead to defining appropriate levels of biodiversity for the long-term sustainability of animal agriculture and aquaculture.

Reproduction Technology

We need research on the reproduction technologies required to underpin breeding and the effective dissemination of genetic improvement to all producers. Artificial insemination is long established in many livestock systems as a central method of animal reproduction with an essential role in breeding programmes and genetic dissemination. More recently, additional techniques have been developed to further improve genetic progress, animal health, and animal welfare, and to disseminate breed improvement. Technologies such as embryo transfer, cloning, and sexing are under development and have had some limited application. Initially, future work on reproduction technologies will aim for a better technical, physiological, and genetic understanding of current and new technologies such as nuclear transfer and cloning for research purposes and semen control of sex determination (e.g. through semen sexing). This will include developing more reliable predictors of semen and embryo quality and studies into the interactions with the environment, nutrition, and infection. Benefits should also stem from research on more reliable indicators of oestrus, conception, and the survival of cryopreserved gametes and embryos. In aquaculture, additional techniques (e.g. for sex determination) and knowledge on the limitations of reproduction are needed.

Future research on cell differentiation may open the way to producing gametes from stem cells. Coupled with predictive biology and statistical techniques such as genome-wide selection, these approaches could make it possible to produce and select multiple generations in the Petri dish. Clearly this entails significant risks as well as potential advantages, so such developments are probably beyond the timescale of our vision.

Reproductive technology research will contribute to safe and wide dissemination of valuable genetic material so as to achieve genetic progress and quick transfer to the production level. Transparency about new technological developments and an open dialogue with society are important in such a sensitive area.

In some livestock systems, such as beef cattle, sheep, goats, horses, buffalo, and most aquaculture species artificial insemination (AI) is still very rare. Reasons for this are many. One of the main reasons is the difficult extensive environment in which these species are usually bred, while AI requires a minimum level of technological input. This applies to a rather large part of livestock production in Europe, and therefore adaptation of the required technology is certainly important.

Biotechnologies

The new cell-based strategies described above have the potential to add to the desired developments in farm animal breeding. More controversial technologies such as genetic modification could be even more powerful, delivering step changes in sustainability in ways unlikely to be possible through conventional selection. For example, there are GM pigs in Canada claimed to have a reduced environmental impact because they have been given a digestive enzyme not naturally present in pigs. This enzyme allows better utilisation of phosphate in the pig's diet, reducing the pollution per unit output. It is debatable whether these pigs will ever be economically viable, but they illustrate the current state of technology. Future research is likely to deliver other GM options, one of the most exciting being the potential to make animals completely resistant to specific viral diseases. If genetic modification (RNA interference) could make chickens resistant to avian influenza and reduce the risk of a human flu pandemic – potentially saving millions of lives – would this be an acceptable use of genetic modification? Despite current public sensitivity, research in the US and Asia is likely to increase pressure on EU agriculture to accommodate novel methods, including the use of nuclear transfer ("cloned") animals for breeding in the pig industry. It is noteworthy in this regard that the US Food and Drug Administration is expected soon to allow products from cloned animals to enter the food chain, with potentially significant implications for international trade, via WTO negotiations for instance. Research in these areas can be part of a medium- and long-term strategy. It can provide information on risks and benefits, enable European society to make informed decisions, and help breeders to define future economic strategies. Research and appropriate communication can also accommodate potential shifts in European public perception in the coming decade. Genetic modification, research on large-animal embryonic stem cells, and novel methods we have not yet thought of will provide valuable research tools for reducing the number of animals used in research.

Plant breeders are already 'stacking' multiple genetic modifications in an individual crop. Animal breeding is some years behind, but targeted gene modifications in large animals are required for controlled experimental work. Future research should thus include a focus on: i) basic research in animal embryology and reproduction, ii) deriving farm animal stem cell lines and tools for controlling the differentiation of these lines, and iii) cloning and gene transfer methods.

The need for this type of research stems from Europe's wish to be a knowledge-based economy while taking into account the ethical climate in Europe. It does not stem from current application requests in the animal breeding industry. Yet to adopt a science-based stance, Europe must support the development of new technologies within its borders. We thus strongly support European and national research into new technologies, also in farm animals, as *basic research,* to be conducted in a transparent way. This includes dialogue with stakeholders and society.

Socio-Economic Research

Society and societal attitudes are constantly evolving and are crucial to defining the ethical, legal, and market context in which animals are produced and animal breeding decisions made. Worldwide and even across the EU-25, there is considerable variation among individuals in their attitudes and behaviours towards animal agriculture and aquaculture. It is consequently very difficult for animal breeders to predict future trends. If animal breeding research is to be relevant and animal selection decisions are to be appropriate, biological research must go hand in hand with socio-economic research shedding light on issues such as segmentation among consumers and how beliefs and behaviours interact.

Here are some of the topics that socio-economic research should address: the role of animal agriculture in sustainable rural economies and rural environments; developing better tools to explore how best to legislate on biotechnology matters and how legislation will affect European competitiveness, consumer health, and food prices; developing effective strategies for communicating information on animal breeding science and technology and for transparency.

Equipped with this vast and growing array of extremely powerful technologies, animal breeding is able to address a great many industrial and societal needs. Yet selective improvement of farm animals must be undertaken in a context of continuing general animal science. As the genetic makeup of farmed animals changes we can expect the nutritional needs of the stock to change as well. This will have an impact on which husbandry practices are most appropriate. Producers with the latest breeding stock also need information and advice on the best available management technologies.

References

Code-EFABAR. C*ode of Good Practice for Farm Animal Breeding and Reproduction* (FOOD-CT-2003-506506). www.code-efabar.org

Genesis Faraday. (2004). *Selective Breeding for Aquaculture – Current Trends and Future Prospects.* Workshop report. http://www.genesis-faraday.org 7pp.

Gjedrem, T. (1997). Selective breeding to improve aquaculture production. *World Aquaculture*. 33-45

Van Genderen, A., de Vriend, H. (1999). The future developments in farm animal breeding and reproduction and their ethical, legal and consumer implications. Report. Project Farm Animal Breeding and Society. EU-QLRT-1999- http://www.effab.info-publications. 130 pp.

5

Animal Healthcare

Livestock production is constrained by a multitude of factors including disease, malnutrition, poor management and ineffective utilization of the best-suited genetic material. The major objectives of animal health activities are to secure food supply for a growing human population, to safeguard human health by combating zoonoses and to facilitate domestic and international trade in animals and animal products. The necessary interventions need to be both technically sound and cost-effective.

During the last two decades spectacular developments in biotechnology, informatics and information systems have been achieved, and the results are now being utilized in the planning and execution of animal disease prevention and control programmes in many countries.

Epidemic Diseases of Livestock

The control of epidemic diseases of livestock has, by definition, a geographical emphasis (country, region, continent), while health and productivity improvement schemes are usually designed for the farmer and/or local community. In the former case, veterinary interventions are most successful if uniformly applied over a prescribed area. In the latter, interventions can vary from farm to farm, depending on clinical conditions and farming practices.

There is a growing desire among veterinary services to assess current disease control programmes and, if possible, move from the control phase to an eradication phase. In order to facilitate this, the International Office of Epizootics (OIE) is defining new guidelines based on a three-stage pathway that a country will need to follow before it can obtain international recognition as being free from an epizootic disease.

For Stage 1 - Provisional freedom from disease - the director of veterinary services will declare this free state to the OIE following the absence of disease in the country and the cessation of vaccination. Stage 2 - Freedom from disease - requires that the country has ceased vaccination for a prescribed minimum period, has had no outbreaks of the disease after declaring provisional freedom and demonstrates results of prescribed surveillance before being internationally recognized. To attain Stage 3 status - Freedom from infection - the country must have extended surveillance after achieving freedom from disease status, have demonstrated evidence of the absence of the causative agent within the livestock population and have in place adequate measures to prevent the reintroduction of infection.

National, subregional and regional eradication will, of course, have beneficial effects for the areas involved. Yet it may not be cost-effective to achieve this, since emergency preparedness to assure the state of freedom is expensive and needs to be maintained continuously at a high level. It can be concluded, therefore, that total eradication gives the best economic return in the long term.

Based on the experience of the FAO European Commission for the Control of Foot-and-Mouth Disease and current activities of the FAO-coordinated Global Rinderpest Eradication Programme (GREP), the following guidelines have been defined as essential components for successful disease eradication:

- coordinated mass vaccination campaigns leading to a verifiable elimination of persistent endemicity;
- use of high-quality internationally recognized vaccines, independently tested for efficacy and safety;
- establishment and proper management of national veterinary services capable of organizing an intensive and sustained control and surveillance programme;
- adherence to the relevant OIE guidelines and time frame for the declaration of freedom from disease and from infection;
- provision of a national laboratory service capable of providing or developing rapid and effective differential diagnostic services;

- articulation of an effective strategy to prevent the reintroduction of the etiological agent;
- development of effective national/regional emergency plans, including a prerehearsed action programme in case of an outbreak. This should include provision for the implementation of a stamping-out policy.

These measures must be sustained by the following legal powers vested in national veterinary authorities:

- compulsory notification of any suspected cases by the owner and/or relevant local authorities;
- authority to collect samples for laboratory investigation;
- powers to enforce compulsory quarantines of infected premises, preferably with slaughter of infected animals, and ring vaccination, and powers of seizure of relevant material;
- payment of compensation;
- sanitation measures and other appropriate procedures on infected premises;
- powers to introduce control of livestock movements and to stop vehicles and herds so that animals may be inspected;
- powers to designate protection and surveillance zones for the purpose of implementing further intensive control measures; powers to implement emergency vaccination campaigns.

Although global or regional eradication of epidemic diseases would be the ultimate achievement, control remains the sole reasonable goal in most situations, either because of the epidemiological complexity, for example, vector-transmitted diseases such as African horse sickness, African swine fever and Rift Valley fever, or because of the inadequacy of available tools to achieve eradication, such as effective and safe vaccines and systems for accurate diagnosis and surveillance coupled with the availability of adequate financial and operational resources.

Eradication at the global or regional level cannot be accomplished without a coordinated international effort in gathering disease intelligence and implementing control procedures. International agencies such as FAO, OIE and the World Health Organization (WHO) have a crucial role to play in this since their mandates enable them to operate; in the international arena relatively free from political constraints.

A major component in the eradication of any major epidemic disease is a viable prevention programme. Realizing this, FAO has initiated a new priority programme called Emergency Prevention System (EMPRES) for Transboundary Animal and Plant Pests and Diseases. The animal-diseases component of EMPRES aims at strengthening the response of member countries in prevention and/or their immediate response to emergencies caused by major transboundary pests and diseases through FAO assistance, including technical coordination of control/eradication activities.

Strategy for Disease Eradication at the Global Level

FAO is coordinating the Global Rinderpest Eradication Programme (GREP) aimed at total eradication of rinderpest from the world by the year 2010. In order to facilitate the execution of this programme, FAO's Animal Production and Health Division has created the GREP Secretariat with the following responsibilities:

- organization of coordinating meetings;
- endorsement of targeted investigations, applied research and global monitoring of viruses by the World Reference Laboratory;
- creation of scientific advisory groups to support the project;
- harmonization of campaigns and risk analysis studies;
- support of global monitoring;
- production of technical guidelines and information/data dissemination.

The GREP Secretariat within FAO will therefore provide the technical linkage between the three regional campaigns in Africa, West Asia and South Asia, which are, respectively, the Pan-African Rinderpest Campaign (PARC), the Western Asia Rinderpest Eradication Campaign (WAREC) and the South Asia Rinderpest Eradication Campaign (SAREC). The success of the regional campaigns ultimately depends on the effectiveness of the operations at the national level, including the veterinary services, the regional infrastructures and the availability of adequate financing. In order to ensure national success, FAO is making technical contributions within each of the regional campaigns in the areas of epidemiology, such as surveillance and sero-monitoring, vaccine quality control and the design of communication programmes to promote community participation.

A Strategy for Disease Eradication at The Continental Level

Foot-and-mouth disease (FMD) is the most important animal disease constraining world trade. In addition, it also directly limits agricultural production through losses in weight gain, milk yield and draught power. The present global distribution of FMD somehow mirrors the world economic order: all Organisation for Economic Cooperation and Development (OECD) countries are free of the disease and operate a policy aimed at total exclusion of the virus. Middle-income countries, such as many South American countries, run major FMD control campaigns to be able to maintain their international trade capability, while in low-income countries, with negligible participation in international trade in animals and animal products (sub-Saharan Africa and some parts of Asia), there are only few examples of effective control programmes.

To a large extent, the success of the ongoing continental plan for the eradication of FMD from the Americas, the Hemispheric Programme for Eradication of Food and Mouth Disease - which has already resulted in several countries in the region attaining disease-free status, can be credited to the national veterinary services and the coordinating role of the South American Commission for the Control of Foot-and-Mouth Disease (COSALFA), together with the technical support of the Pan-American Foot-and-Mouth Disease Center (PANAFTOSA). This illustrates the effectiveness of international collaboration and should serve as a model for other developing countries engaged in regional disease control/eradication efforts. It is befitting that an in-depth long-range review of FMD eradication strategies in the next millennium will take place in South America, where the first Joint FAO/OIE/Pan-American Health Organization (PAHO) Conference on the Perspective for the Eradication of Foot and Mouth Disease in the Next Millennium and the Impact on Food Security and World Trade will be convened in 1996, most probably in Brazil.

Control of Endemic Diseases

Brucellosis

Brucellosis continues to be a problem in many countries and it is the most important zoonotic disease in the Near East. FAO, together with WHO and OIE, is preparing project documents for a regional brucellosis control programme according to guidelines established in collaboration with the countries of the region. The strategy of the programme is based on whole herd/flock vaccination of cattle, buffaloes, sheep, goats and camels regardless of age or sex. It is expected that after a period of 15 to 20 years of mass vaccination, the incidence of the disease will be brought to a level that will make the concept of disease eradication realistic. In the meantime, improved tools to fight the disease - a more potent vaccine and/or serological tests that can differentiate between infected and vaccinated animals - will have been made available to match the new goal.

Ticks and Tick-Borne Diseases

The strategy of the FAO ticks and tick-borne diseases programme aims at promoting integrated tick and tick-borne disease control methods that include immunization when applicable and increasing awareness of the fact that tick acaricide resistance is spreading worldwide.

This approach has been the philosophy behind the Coordinated Programme for the Control of Ticks and Tick-borne Diseases in Central, East and Southern Africa currently entering its third phase. The programme's objective is the control of ticks and tick-borne diseases in cattle based on immunization against theileriosis, heart-water, anaplasmosis and babesiosis on a sustainable cost-recovery basis, coupled with strategic tick control.

To complement this approach, FAO supports work on the assessment of acaricide resistance on a worldwide basis in collaboration with the World Acaricide Resistance Reference Centre in Berlin, Germany, and through the organization of workshops such as those held in Zimbabwe, Malawi and Brazil in 1994.

Tick eradication is considered when special circumstances facilitate this concept, such as in the case of *Amblyomma variegatum*infestation in the Caribbean. After more than seven years of studies and negotiations, the Programme for the Eradication of *Amblyomma variegatum* and Its Associated Diseases, Heart-water and Dermatophilosis, from the Caribbean has been declared operational. This followed agreements between FAO and the Inter-American Institute for Cooperation on Agriculture (IICA) and FAO and the Caribbean Community Secretariat (CARICOM).

The programme has now begun, and work is in progress to map the occurrence of the tick through carefully designed surveys. A pilot eradication programme including maximum community participation has been launched in Anguilla, and this will be followed by the planning and execution of a large-scale eradication programme aimed at eliminating this serious pest from the region before the year 2000. The start of the programme is very timely as new foci of *A. variegatum* infestation have recently been reported from Dominica and new areas in Barbados.

Insect-borne Diseases

African animal trypanosomiasis (AAT), transmitted by tsetse flies, is arguably the most important animal health constraint in sub-Saharan Africa. FAO, through its Programme for the Control of African Animal Trypanosomiasis and Related Development, provides the coordination required for a concerted approach to this complex, multifaceted problem. The FAO group dealing with insect-borne diseases has established a Geographical Information System (GIS) to quantify the economic and agricultural impact of AAT and to identify the areas where AAT control is most likely to translate into increased agricultural productivity. The disease is gradually assuming more importance in the semi-arid and subhumid areas. Control efforts are particularly successful in areas of expanding agriculture. A noteworthy development is the availability of novel bait techniques such as traps, targets and insecticide-treated livestock. FAO's contributions in this regard involve support to training, applied research and education of rural community groups.

Non-tsetse-transmitted animal trypanosomiasis (NTTAT) has a much wider geographical distribution and comprises the mechanically transmitted *Trypanosomiasis evansi*, or Surra, the sexually transmitted *Trypanosomiasis equiperdum*, otherwise known as Dourine, and the non-cyclical transmitted *Trypanosomiasis vivax*. The situation with regard to the distribution of *T. evansi* is also dynamic. Transboundary movement of humans and animals is believed to be associated with recent flare-ups of Surra in cattle in Indonesia and in carabaos, the swamp buffalo, in the Philippines. The latter outbreaks are being investigated under the FAO Field Programme.

With the successful conclusion of the screw worm eradication campaign in the Libyan Arab Jamahiriya supported by the Screw worm Emergency Centre for North Africa (SECNA) - which involved the active participation and collaboration of the Mexico/United States Commission on Screw worm - FAO has shifted attention towards the prevention of new outbreaks of both New World Screw worm (NWS) and Old World Screw worm (OWS), as well as of other exotic diseases. Training activities have so far mainly concentrated on North Africa. A proposal for NWS eradication in Jamaica has been submitted for donor funding. It is envisaged that this programme will lead to a large-scale eradication campaign in the Caribbean. Activities continue to improve techniques for NWS surveillance, control and the use of the Sterile Insect Technique (SIT) for eradication, in the anticipation that the results will yield significant benefits to all infested countries.

Non-Infectious and Production Diseases

To a large extent, the great effort to control the major infectious and parasitic diseases has been successful in many parts of the world. It has not always resulted in the expected increase in livestock production and productivity, however. The reasons for this are many and varied, including a number of important production-related diseases, such as helminth infections, as well as reproductive disorders, nutritional factors and other non-infectious conditions that have been neglected. These are only part of a complex set of circumstances - other factors may include the lack of government incentives, erroneous price policies and a failure to deliver animal health service - that impede production.

During the last few years FAO has put more emphasis on the development of programmes for the control of helminths and non-infectious diseases. A number of activities have been aimed at increasing governments' awareness of the economic importance of these diseases and conditions and the constraints they put on efforts to accelerate livestock productivity. Activities include the publishing of informative materials and the generation of production data through pilot projects.

The rapidly increasing problem of anthelmintic resistance in sheep parasites is of great concern to FAO, and funds have been allocated to activities attempting to map the extent of the problem in developing countries. Through a recent survey supported by the FAO Technical Cooperation Programme (TCP), the situation in the southern part of Latin America has been evaluated. The results are critical, with more than 75 percent of farms showing problems of resistance to two of the three main groups of anthelmintics currently available for chemical control and a growing number showing resistance to all three. A consultancy to evaluate the situation in selected African countries has pointed out that similar problems exist in eastern and southern Africa.

Parasitic zoonotic diseases (hydatidosis, cysticercosis, trichinellosis) continue to be a major problem causing human suffering and significant losses from condemned meat and organs. FAO, in collaboration with the Veterinary Public Health Units of WHO and PAHO, is developing control strategies, and TCP has supported the evaluation of current programmes in some South American countries.

Micromineral deficiencies affecting animals (P. Cu. Co, Zn, Cu-Mo-S interactions) are well known in many districts of the world. They may greatly reduce animal productivity and food security for livestock owners and consumers alike. Because of the unavailability of mineral supplements or fertilizers, however, control cannot always be achieved, even though the knowledge for control is available.

Animal Health Delivery at Herd Level

FAO's livestock programme is increasingly emphasizing an integrated approach to livestock production and enhancement of food security. This strategy aims at focusing much technical and social expertise on priority livestock systems. Two such systems are mixed farming (crop and livestock) in high potential areas and peri-urban dairy production.

One arm of this integrated approach is a Herd Health and Production Programme (HH&PP) protocol, which includes the following interactive steps for delivering animal health within farming systems:

- agree with farm management on proposed production and acceptable risk targets;
- define and develop the tools and technologies required to collect the necessary data;
- interpret the data gathered from all sources;
- develop and agree upon a plan of action that includes concrete steps to reach the production and acceptable risk targets set in step one;
- monitor and evaluate progress and begin again at step one.

This approach allows private veterinary practice to deliver clinical and HH&PP services and defines areas of responsibility in a manner that promotes close cooperation between the private sector and public services through the implementation pilot programmes using the HH&PP protocol and by closely monitoring results. In addition, national consultative groups of interested persons for HH&PPs can be established to define further the concept and its delivery, to modify the approach to local conditions and to disseminate results. These consultative groups would include government and private-sector providers, producers of livestock inputs, educators and other interested parties.

Enabling Domains

The key tools for the international control of animal diseases are the availability of adequate information, proper services for the diagnosis of disease conditions, vaccines of appropriate quality and a functional veterinary service. Emerging trends in all four areas will affect future strategies for the control of animal diseases.

Information

The international exchange of information concerning animal diseases is coordinated by OIE, which classifies infectious diseases into List A and List B, reflecting the contagiousness of the disease and its relative economic importance for international trade in animals and animal products. It is the responsibility of the Chief Veterinary Officer of each country to notify the international community, through the OIE, whenever an outbreak of an epizootic disease is encountered and periodically to update OIE on the disease situation in the country. Every year FAO publishes the FAO/OIE/WHO *Animal Health Yearbook*, which details the disease status of each member country of OIE, WHO and FAO.

The three international organizations have recently been evaluating the quality and impact of the information contained in the yearbook, and it is widely acknowledged that the official information given in the yearbook often represents the minimum known about the disease situation in many developing countries. Therefore, it is now well recognized that veterinary services need to develop sound information systems that would strengthen the services and improve the accuracy of disease incidence data and of diagnostic testing or surveillance results. This would also enable governments to devise appropriate livestock health and productivity improvement schemes at community level. OIE, together with FAO, the Inter-American Institute for Cooperation on Agriculture (IICA) and other agencies, has commissioned the development of such a computer program, HANDISTATUS. The objective of FAO and OIE is to provide a system that would greatly improve such an exchange of disease information, enabling each national veterinary service to be readily aware of the disease status of every member country of OIE and/or FAO. The international challenge will be to ensure that developing countries have access to the system and use it effectively.

A similar comprehensive disease-notification system for vesicular diseases has been developed for South America by PANAFTOSA, which, together with IICA and others, is currently developing various information systems. In this connection, the project "Prevención de las principales enfermedades exóticas de los animales en América Latina y el Caribe" (PROPEXAN) has a crucial complementary role. This project has been operating mainly in Central America, the Caribbean and Andean region countries. Based in Panama, it offers formal and practical in-service training to officers from countries in the region on animal health inspection, animal quarantine and the epidemiology, prevention, identification and eradication of exotic animal diseases.

Diagnosis

Current trends in animal disease diagnostic technology are moving towards an improved flow of samples and data from herd level to the veterinary clinic, national laboratories, regional reference laboratories and world reference laboratories. While conventional techniques in veterinary diagnosis should not be overlooked, the use of modern diagnostic technology based on molecular biotechnology is dramatically gaining ground, and polymerase chain reaction (PCR), monoclonal antibody probes, enzyme-linked immunosorbent assays are now becoming standard, primary diagnostic tools in industrialized countries. There is a need for technology transfer, however, in order to establish these technologies in developing countries.

FAO has been addressing this issue from four aspects:

- Training and awareness of decision-makers. The approach has been through a series of awareness workshops directed at senior scientists in developing countries, enabling them to appreciate the potential offered by the new technologies in animal health. Where a defined role has been identified, FAO, either alone or with other organizations such as the International Atomic Energy Agency (IAEA), has convened special expert consultations to advise on the specific subject matter.
- Practical training of technicians and laboratory scientists. This has taken the form of training at the bench in specific techniques for specific

purposes. The most extensive programme has been the training of personnel from developing countries in the enzyme-linked immunosorbent assay (ELISA) technology conducted by the Joint FAO/ IAEA Division. Through this initiative, it has been possible to develop surveillance programmes such as the rinderpest sero-monitoring network for GREP and the FMD sero-monitoring network that is being developed in South America in collaboration with PANAFTOSA.

- FAO biotechnology networks. These are currently being developed in eastern Europe, India and South America. The activities in South America are primarily geared towards strengthening the regional capability of diagnosing those diseases that are not yet covered by the excellent programme of PAHO. Thus, the FAO Regional Office in Latin America and the Caribbean has been developing a network of leading laboratories, called REDLAB, in order to promote the exchange of expertise, information and reagents through a system of TCDC (Technical Cooperation among Developing Countries).
- Global monitoring of pathogens of major diseases. This is being achieved through a network of reference laboratories selected on the basis of their established expertise in providing referral diagnosis for the disease(s) for which the laboratory is designated. For example, for FMD, FAO has designated the Pirbright Laboratory of the United Kingdom's Institute for Animal Health as the FAO World Reference Laboratory, while PANAFTOSA is the FAO Regional Reference Laboratory for the Americas. Similarly, FAO has just designated the Pirbright Laboratory as the FAO World Reference Laboratory for Rinderpest to support GREP. The data being generated by the network of reference laboratories are proving to be of crucial importance in molecular epidemiology and global disease monitoring. The networks of FAO, OIE and WHO complement each other.

Vaccines

Recent trends in vaccine development have involved three principal issues: vaccine technology, vaccine quality and vaccination monitoring.

Developments in vaccine technology with an immediate impact have been associated with the consolidation of cell culture and fermentation technologies, gene deletion, gene manipulation of vaccine viruses and viral-vector recombinant vaccines, while subunit genetically engineered or synthetic peptide vaccines, so far, have not yielded spectacular results. FAO hat published a comprehensive review of and guidelines for veterinary vaccines.

Perceptions and attitudes regarding vaccine quality have also been evolving. It is now widely accepted by most developing countries that vaccines should be manufactured under conditions of good manufacturing practices (GMP). The concept that FAO currently promotes is one that was originally provided by the United Kingdom Veterinary Medicines Directorate and that was endorsed by the FAO Expert Consultation on Quality Control of Veterinary Vaccines in Developing Countries held in 1991, namely:

> "Medicines (vaccines) must be manufactured with appropriate quality control procedures in premises that are inspected and licensed; the ingredients must be of appropriate purity, in correct proportions and correctly processed; the containers must be robust with secure closure; the labelling must be accurate and informative."

In Africa, these concepts of vaccine quality have been championed by the Pan African Veterinary Vaccine Centre (PANVAC), which is a programme of the Organization of African Unity/Interafrican Bureau for Animal Resources (OAU/IBAR) with technical assistance from FAO. The greatest impact of PANVAC has been in the improvement of the quality of rinderpest vaccine, as evidenced by the fact that the proportion of African-produced vaccine lots meeting international standards of quality has risen from 33 percent in 1985 to about 80 percent in 1992. Also, the quality of testing has reached international standards as reflected by the diversity of causes for vaccine rejection.

In Latin America and the Caribbean, PANAFTOSA/PAHO has promoted and trained technical personnel in the production, quality control and use of FMD vaccines. The Pan-American Institute for Food Protection and Zoonoses (INPPAZ)/PAHO has played an important role in training in quality control and production of vaccines against rabies and brucellosis.

The advent of ELISA technology has made sero-monitoring a practical reality for vaccination monitoring. The Pan-African Rinderpest Campaign (PARC) sero-monitoring network coordinated by the Joint FAO/IAEA Division is the most extensive sero-monitoring network in the world. The crucial elements of this network are the application of the ELISA technique using standardized and highly specific reagents, standardized equipment and computer software, supported by a well-controlled quality-assurance programme for the reagents and techniques by consistent technical advice from FAO/IAEA personnel and the scientist responsible for developing and standardizing the technique. A standardized training programme is also included for all individuals responsible for the national testing programme. It is now FAO policy to extend the approach to other parts of the world involved in GREP. A similar scheme is being developed by the Joint FAO/IAEA Division to monitor the effectiveness of FMD and brucellosis control programmes in South America.

Veterinary Services

A strong veterinary service with a clearly defined chain of management and reporting command is a prerequisite for effective control of animal diseases. The veterinary service should also have adequate laboratory support for disease diagnosis and surveillance. The national Chief Veterinary Officer (or Director of Veterinary Services) should have a direct line of functional command to the laboratory regarding responsiveness to priority contagious diseases and aspects of their control requiring laboratory support.

An effective veterinary service does not necessarily mean government action in all veterinary matters. Schemes for fostering a healthy private veterinary programme need to be strongly supported by national veterinary associations and international organizations. From the point of view of controlling animal diseases, it should be noted that a disease outbreak is initially encountered by the farmer, herder or pastoralist or those responsible for on-farm routine animal health activities, normally private veterinary personnel and/or community-based extension workers. Structures that encourage licensing of private veterinarians to undertake such activities as routine vaccinations, on-farm testing and even clinical inspections need to be developed. Such licensed veterinary inspectors should be adequately remunerated by the government for activities subcontracted to them, however, and they should be allowed to charge fair fees for services rendered.

Control of animal diseases demands that national veterinary services operate within both a regional and an international context. An outbreak of a major epizootic disease in one country is of immediate concern to neighbouring countries and should be reported to OIE and FAO, and, in the case of zoonotic diseases, to WHO as well. To be effective in controlling epizootic diseases, each country within a region must have a strong national veterinary service that freely exchanges information about disease with other countries. And as control of epizootic diseases must often take "public interest" into account more than just "farmer interest", services are largely government financed and directed.

It is evident from the trends in the international control of epizootic diseases that countries, particularly developing countries, will need to strengthen and modernize their veterinary services and not regard them simply as a mere component of an agricultural extension service.

Prevention of Transboundary Animal Diseases

Fifty years ago, FAO held its inaugural conference in Quebec, Canada, and, within a year, a meeting on animal health was convened in London, the United Kingdom, to consider how the activities of veterinary organizations all over the world could be properly coordinated under the FAO umbrella, with particular attention being paid to mitigating the widespread ravages of animal plagues, especially rinderpest.

The following year, the subcommittee on animal health of the FAO Standing Advisory Committee on Agriculture recommended that FAO should assist in the distribution and establishment of novel rinderpest vaccines. Thus, the coordination of international efforts to control major epizootic diseases became part of FAO's basic mandate, which has been pursued over the past 50 years with varying levels of emphasis.

In February 1994, Dr Jacques Diouf, the incoming Director-General of FAO, pledged to remodel the Organization to take a leading role in global food security, especially in low-income food-deficit countries. It was therefore not surprising when he identified resolute action against major transboundary epizootic diseases and migratory plant pests as complementary components of a priority programme. Consequently, the Director-General sought and obtained the mandate of the 106th FAO Council, in June 1994, to establish a priority programme - the Emergency Prevention System for Transboundary Animal and Plant Pests and Diseases - to be known by the acronym EMPRES.

The livestock diseases component of EMPRES aims at strengthening FAO's role in the prevention of, and immediate response to, emergencies caused by major epizootic diseases of transboundary importance. The primary thrust of this component is against rinderpest and the focus is the Global Rinderpest Eradication Programme (GREP). Both the Director-General and the FAO Council intend that EMPRES shall cover those transboundary diseases and pests that constrain food security, adversely affect public health or impede international trade in livestock and animal products. Transboundary animal diseases are defined as epizootic diseases that are highly contagious or transmissible with the potential for very serious and rapid spread, irrespective of national borders.

Through EMPRES, the FAO Council has presented the Organization with a major challenge. FAO is expected to provide technical leadership and coordination for GREP as well as for an international effort in the prevention, prediction and containment/control of emergencies caused by a defined list of major epizootic diseases, such as rinderpest, foot-and-mouth disease (FMD), Peste des Petits Ruminants (PPR), contagious bovine pleuropneumonia (CBPP), Rift Valley Fever and lumpy skin disease. The challenge is that these are daunting tasks of almost unprecedented magnitude. No animal disease has ever been eradicated from the world in a manner akin to the global eradication of smallpox, and there is no precedent for effecting international coordination of emergency disease prevention. The opportunity is that both tasks represent a major vote of confidence in the technical capability and unique disposition of FAO to lead the world in this vital area, bearing in mind the impact of the identified diseases on world food security and on international trade in animals and animal products.

The Director-General has given clear guidance that the strategy for EMPRES should be within the overall context of its prime elements: early warning, early reaction and enabling research. The goals, objectives and strategy set by the Director-General have been fully endorsed by the FAO Council. Accordingly, during its review of the initiation of the implementation of the Director-General's Special Programmes, the 107th FAO Council, in November 1994, remarked:

> "In respect of... EMPRES, the Council noted the ongoing activities to strengthen FAO support to global rinderpest eradication and expressed its appreciation of the emergency assistance provided to several countries in Africa linked to risks of serious epidemics of rinderpest and contagious bovine pleuropneumonia. In this connection, it underscored the intention to also address other important animal diseases, such as foot-, and-mouth disease...."

As Gordon Scott and Provost remarked, rinderpest belongs to a select group of notorious infectious diseases that have changed the course of history. It is the most dreaded bovine plague known. Rinderpest is a highly contagious and lethal viral disease of cloven-hoofed animals, both domesticated and wild. Cattle and buffaloes are particularly affected, but wildlife, such as the African buffalo, antelopes, wild pigs and giraffes, are also affected severely when the disease is introduced by cattle. Typically, rinderpest causes fever, eye and nose discharges and necrosis of the lining of the digestive tract, giving rise to severe bloody diarrhoea and dehydration, followed by death or protracted convalescence. It is acknowledged to be the world's most damaging virus infection affecting farm livestock, causing reduced meat and milk production as well as reduced crop production through loss of draught power. In the past, cattle plague - as epidemic rinderpest was called - destroyed whole populations of cattle, resulting in widespread famines and profound economic and political damage. Organized veterinary education and state veterinary services were established in Europe primarily to deal with eradication of cattle plague during the eighteenth century, when repeated invasions came from the East. In later years, rinderpest provided the stimulus for the establishment of state veterinary services in other continents. Indeed, as already remarked, one of the main reasons for the establishment of FAO itself was the need for international coordination in the fight against rinderpest.

Experience has shown that regional coordination of campaigns is the single realistic approach to rinderpest control; isolated national actions have only led to sporadic and unsustained improvements. Stringent control of livestock movements, slaughter of affected herds and strict quarantine of the infected zones are the principal elements of rinderpest eradication. The discovery of effective vaccines in the 1940s, and, more recently, the introduction of cell culture vaccine in the 1960s, facilitated the organization of large-scale rinderpest control campaigns, such as the All-India National Campaign and the Joint Project 15 programme in Africa, with remarkable reductions in the incidence of rinderpest. Unfortunately, such dramatic results were not always associated with the total elimination of rinderpest from the affected region. Consequently, several years later, epidemics of rinderpest reappeared in India, Africa and the Near East as a result of the introduction of animals infected by residual endemic foci into a susceptible population. This new rinderpest pandemic in the early 1980s, once again, killed millions of bovines (cattle and buffaloes) as well as wildlife, and thousands of farmers and herders lost most or all of their herds. Nigeria alone lost an estimated one million cattle, plus wildlife. In Iraq, at least 18 000 buffaloes and cattle died. Estimates by FAO suggested that the direct losses resulting from the 1980-1984 rinderpest epizootic in Africa alone were US$400 million to $500 million and indirect losses reached US$1 billion. The recent introduction of rinderpest into the Northern Areas of Pakistan resulted in a major epizootic, killing more than 20 000 buffaloes, cattle and yaks in 1994.

Recognizing the continuing impact of rinderpest on food security and the shortcomings of periodic rinderpest control campaigns, the FAO Expert Consultation on the Strategy for Global Rinderpest Eradication, held in October 1992, concluded that total eradication of rinderpest infection from the world was economically justified and technically feasible. Global eradication was considered to be the only viable option available to overcome cyclic rinderpest emergencies that cause heavy economic and social losses, thereby necessitating international intervention to finance expensive campaigns. FAO was urged to collaborate with other interested organizations, including the International Office of Epizootics (OIE), in order to promote and coordinate a time-bound programme aimed at eradicating rinderpest from the world by the year 2010, to be known as the Global Rinderpest Eradication Programme (GREP).

Global Rinderpest Status and Eradication Measures

The eighteenth century saw rinderpest extend its range from central Asia into Europe, where the cattle plague caused widespread economic and political damage. During the present century, Europe has been essentially free of the disease except for minor rapidly controlled outbreaks in Belgium, in 1920, and in Georgia, in 1990/91. Only one incursion into the Americas is known and the continent remains free of rinderpest today, as does Oceania. The extreme east of Asia has been free of the disease for some time. Despite successful control in the more developed world, however, extensive areas of the globe remain affected. Four main interlinked theatres are recognized. The understanding presented here is a synthesis of information derived from many sources and includes epidemiological insights generated from missions undertaken by EMPRES and related projects.

Africa

After the disease's introduction into northeastern Africa at the end of the nineteenth century, a rinderpest pandemic devastated both domesticated and wild ruminant populations throughout the entire continent. Following the initial virgin pandemic, the disease persisted on the continent, giving rise to repeated cycles of mass destruction such as that evident in the 1970s and 1980s following the rinderpest control campaign Joint Project 15. Although this campaign almost achieved continental eradication, two small endemic foci remained in West Africa, at the border between Mali and Mauritania and in Ethiopia.

The resumption of the disease resulting from these foci brought about the need for yet another campaign. The present ongoing internationally coordinated control programme - the Pan-African Rinderpest Campaign (PARC) - although again successful in freeing large areas of the continent, has not yet achieved eradication. Currently, the disease is recognized only in eastern Africa, persisting in discrete endemic foci where it may not be manifested dramatically. These endemic foci, associated with the large cattle herds on which pastoralist communities are dependent and reaching across international boundaries, have been defined in the northeast of Ethiopia, arriving at the Eritrean border, and in the south of the Sudan, extending into contiguous areas of Ethiopia, Uganda and Kenya.

In addition, an endemic focus has long been suspected to be present in southeastern Ethiopia, and recent evidence suggests that it might even extend into northeastern Kenya and the extreme south of Somalia. From these foci of persistence, epidemics spread periodically into neighbouring areas, primarily through trade in cattle, as occurred in late 1994, for example, when outbreaks of rinderpest resulted in the highlands of northern Ethiopia because of trade in draught oxen from the endemically infected area of the adjacent lowlands.

The Arabian Peninsula

Rinderpest was present in Yemen for many years despite annual vaccination campaigns. It was assumed that infection there was introduced repeatedly by trade in livestock from Africa. However, recent molecular analysis of an isolate of rinderpest virus from that country seems to indicate the persistence of infection with a virus related to the Asiatic rinderpest complex. Elsewhere in the Persian Gulf States of the Arabian Peninsula, sporadic outbreaks appear to result from repeated introduction of infection via cattle traded from South Asia.

Western Asia

Historically, a pattern of waves of rinderpest infection resulting from cattle movements from the East could be detected in countries of the Near East region, but, in recent years, the disease has become established and has persisted in northeastern Iraq, northwestern Iran and southeastern Turkey. Recent emergency assistance by FAO, enhancing national control programmes, has reduced the incidence and might even have eliminated this focus. Until recently, rinderpest persisted in the centre and south of Iraq, but now the disease appears to be limited to the marshes on the channel of Shatt-al-Arab. Spread ostensibly from the northern focus to the very borders of Europe, rinderpest has occurred in Turkey twice in the last five years, but the Caucasus region appears to have remained free of the disease.

South Asia

Once widespread in India, intensive and recent sustained control has restricted rinderpest to the extreme south in the state of Tamil Nadu. Sri Lanka was free from the disease for 47 years, until it was reintroduced in 1987, and it is now persisting in the northeast of the island. In 1993, Pakistan officially acknowledged the presence of rinderpest for the first time in 20 years; samples from the Landhi Dairy Colony near Karachi in Sindh Province submitted to the FAO World Reference Laboratory for Rinderpest (WRL-RP) confirmed the existence of the disease. In early 1994, rinderpest spread to Punjab Province, and later to the Northern Areas. Pathological specimens examined at WRL-RP for identification and genetic analysis validated the close relationship of the rinderpest viruses present in all three areas of Pakistan, confirming the suspected pattern of spread. By the end of 1994, the disease situation in the Northern Areas had deteriorated to a point where mortality of cattle, yaks and buffaloes exceeded 20 000; although still present, it is being brought under control with FAO assistance. Rinderpest has not been recognized in recent years in the rest of Asia with the exception of the Tuva Autonomous Soviet Socialist Republic and Mongolia, where a single epidemic was documented in 1991/92. This raises suspicions concerning the rinderpest status of the central and southern Asian republics.

Guidelines for Global Rinderpest Eradication

OIE has adopted a three-stage pathway that a country must to follow before it can be recognized as being free from rinderpest. Thus, eradication programmes can begin with mass vaccination campaigns within epidemiologically defined zones during which time the veterinary services are also strengthened to prepare them for surveillance activities. Following completion of the campaigns, the country may declare itself "provisionally free from disease", stopping all vaccination activity and replacing it with extremely vigilant surveillance. When a provisionally free country reports no clinical disease for at least three years after stopping vaccination, then, following external verification, it will be declared "free from disease". A further two years after this, if recommended procedures have failed to detect any antibodies in unvaccinated livestock, and again following external verification, the country will be declared "Free from infection with rinderpest virus". In this way, this so-called OIE pathway provides operational targets for GREP.

Based on this pathway, FAO has issued implementation guidelines for regional rinderpest campaigns. Generally referred to as GREP Guidelines, these can be summarized as follows:

- coordinated mass vaccination campaigns to eliminate persistent endemicity;
- the use of high-quality vaccines, independently tested for efficacy and safety;
- proper management of a national veterinary service capable of organizing an intensive and sustained surveillance programme;
- development of a time-bound programme conforming to the relevant OIE guidelines for declaration of freedom from disease and from infection;
- provision of a national laboratory service capable of providing or developing rapid and effective differential diagnosis results;
- articulation of an effective strategy to prevent the reintroduction of the etiological agent;
- development of effective national/regional emergency plans, including a prerehearsed action programme in case of an outbreak that should include provision for a stamping-out policy.

National veterinary authorities should also hold adequate legal powers, including:

- compulsory notification of any suspected cases by the owner and/or relevant local authorities;

- authority to collect samples for laboratory investigation;
- powers of seizure, compulsory enforceable quarantines of infected premises - preferably involving the slaughter and destruction of infected animals - and ring vaccination;
- payment of compensation;
- sanitation measures and other appropriate procedures on infected premises;
- powers to stop vehicles and herds in order to inspect animals;
- powers to designate protection and surveillance zones for the purpose of implementing further intensive control measures;
- powers to implement emergency vaccination campaigns.

Global Rinderpest Campaigns

GREP is being pursued through individual country projects operating under the umbrella of regional coordination units. At present, the European Union (EU)-funded Pan-African Rinderpest Campaign (PARC), coordinated by the Organization of African Unity/Interafrican Bureau for Animal Resources (OAU/IBAR), is fully functional. In South Asia, Bhutan has declared a status of "Provisional freedom from rinderpest" according to the OIE pathway and the Indian National Rinderpest Campaign has scored remarkable success, from having started with a status of rinderpest endemicity in almost all 23 states of the union to the present, where active rinderpest now seems to be confined to only one state and with the northern states advancing towards cessation of vaccinations. The need for regional coordination has now become urgent. Consequently, a proposal for a South Asia Rinderpest Eradication Campaign (SAREC) has been drawn up jointly by FAO, the EU and the countries of the region to commence operating by the end of 1995. A strategy to address western Asia is under preparation through EMPRES. It is expected that this will take into account the current rinderpest epidemiological situation in the region as well as follow up on the previous project that had coordinated rinderpest control in parts of the region.

Within the regional rinderpest campaigns, FAO intends to execute, where appropriate, those elements that relate to either analysis or monitoring, namely epidemiology including sero-monitoring (through the Joint FAD/IAEA Division of Nuclear and Related Techniques in Food and Agriculture), vaccine quality control, communication and project coordination. Funding has now been approved by the EU for such projects in support of PARC through OAU/IBAR. It is intended that the proposed SAREC and other regional projects will have similar components, supplementing FAO-executed coordination projects. Until 1993, these FAO activities were funded by the United Nations Development Programme (UNDP), trust funds and FAO's own Technical Cooperation Programme (TCP) in Africa and western Asia.

A fundamental principle of GREP, promoted through all regional programmes, is the timely withdrawal of mass vaccination campaigns followed by entry on to the OIE-defined pathway. EMPRES supports the goal of eradication by undertaking continuous epidemiological surveillance and by providing immediate assistance with expertise and control requirements. FAO is uniquely placed to do this because of the EMPRES mechanism for rapid response and its ability to follow this up with TCP assistance.

Global Rinderpest Diagnosis and Surveillance

An initial priority for EMPRES was to identify a laboratory capable of acting as the FAO World Reference Laboratory for Rinderpest (WRL-RP), bearing in mind the requirement for a strong presence in molecular biology. Of the several candidate institutes, the Institute of Animal Health, Pirbright Laboratory, United Kingdom, was selected in recognition of its world leadership in rinderpest biotechnology, especially molecular epidemiology, and because it is uniquely placed to receive and process infectious material containing the major epidemic disease agents. The Pirbright Laboratory has undertaken to receive samples from all FAO member countries and analyse rinderpest virus strains using conventional and state-of-the-art molecular techniques, thus complementing field epidemiological/surveillance activities. Through EMPRES, action is being taken to develop a network of strategically sited regional reference laboratories throughout the GREP region with WRL-RP as the central reference source providing expertise, reagents and support. This has the twin aims of enhancing the global rinderpest surveillance capability in developing countries and of disseminating the biotechnology essential for molecular epidemiology.

The FAO WRL-RP uses amplification of viral nucleic acid by the reverse transcription polymerase chain reaction (PCR) followed by hybridization with cDNA probes to provide a highly specific assay that can be used directly on field samples to detect viral nucleic acid with a sensitivity much higher than conventional techniques. Positive results have been obtained where conventional tests failed to disclose the presence of rinderpest virus. Nucleotide sequencing of the F gene of rinderpest virus in amplified products then provides a "fingerprint" of the virus that defines the relatedness of viral strains and, hence, their origins. Global monitoring of rinderpest virus strains through molecular analysis is already proving to be a valuable epidemiological tool to assist in clarifying patterns of disease occurrence and the origins of outbreaks, providing information essential for the elaboration of sound eradication strategies. This approach has proved its worth in confirming the diagnosis of rinderpest in livestock in Pakistan and in Cape buffaloes in the Tsavo National Park of Kenya. It is also providing invaluable assistance in clarifying the relationships between rinderpest outbreaks and endemic foci in East Africa and Asia.

Surveillance under EMPRES is being developed as a multifaceted endeavour primarily to assist countries under GREP but also as part of a global early warning system. All regional rinderpest campaigns are being advised to incorporate a strong element of surveillance, ranging from clinical antibody monitoring using serological surveys, epidemiological disease information systems and molecular analysis of virus strains. The contribution of the Joint FAD/IAEA Division needs to be acknowledged as having resulted in the most extensive sero-monitoring network for any infectious disease, human or animal. Furthermore, it is now planned to extend the surveillance and early warning system to include the application of modern biotechnology, satellite imagery, the geographical information system (GIS), disease intelligence and computer-based decision support systems.

Improving Vaccine Quality

A major recommendation of the FAO expert consultations on rinderpest has been the standardization of the vaccine for use in control campaigns. The FAO Expert Consultation on Rinderpest Diagnosis and Vaccine Production and Quality Control recommended, as a first step for rinderpest campaigns, the use of cell culture vaccine from the RBOK strain developed in the early 1960s by British scientist Dr Plowright at the then East African Veterinary Research Organization, Muguga, Kenya. Further developments in rinderpest vaccine technology seem to indicate that, while the conventional "Plowright vaccine" will continue to be the standard vaccine, the thermostable variant should be preferred for areas that are difficult to access.

Ensuring the quality of vaccines used in national campaigns has been a major FAO contribution to such campaigns. This system has been most extensively elaborated in Africa through the Pan African Veterinary Vaccine Centre (PANVAC). Between 1988 and 1993, PANVAC tested a total of 694 rinderpest vaccine batches for use in PARC. The system of independent quality control is also being extended to campaigns in western and South Asia.

Additional inputs by EMPRES are being applied both to production technology and to the technology as well as the concepts of vaccine quality. In this regard, EMPRES has been promoting controlled trials of novel candidate vaccines from biotechnological developments such as the vaccinia-recombinant and the capripox-recombinant vaccines. It also hopes to encourage the use of modern biotechnology in the quality assessment of vaccines.

Emergency Preparedness Against Epizootics

The EMPRES programme emphasizes the need for efforts to pre-empt disease emergencies through enhanced early warning, early reaction and exploitation of appropriate research. Contingency planning is an important part of emergency preparedness. To this end, a series of workshops on contingency planning and emergency preparedness are currently being held in key regions so as to raise the capacity of countries to cope rapidly with epizootic disease incursions. These deal primarily with the veterinary service structures and preparations needed in case of invasion of countries or regions by major epizootic diseases and with contingency planning for the introduction of major epizootic diseases, using rinderpest, FMD and CBPP as examples to illustrate the concepts. Because of the importance of rinderpest, the workshops in countries under GREP aim to assist those countries that are in a position to cease mass vaccination and progress towards declarations of freedom from rinderpest.

Development of multimedia computer-based communication technology for the production of training manuals and an interactive communication network for the global rinderpest database and global rinderpest surveillance is being undertaken in collaboration with a specialized company together with WRL-RP.

The principles of early warning and early reaction are also being applied progressively to the other five major epizootic diseases addressed under EMPRES as causes of disease emergencies. The recent extension of CBPP to previously unaffected areas of eastern and southern Africa has been of particular concern. FAO's response, thanks to the synergistic facility of EMPRES and TCP, has been rapid and well appreciated by the countries in the region. This CBPP epidemic has been of major concern not only to FAO and the countries concerned, but also to OAU/IBAR, OIE and the EU. Consequently, a regional workshop on the subject, sponsored jointly by EMPRES, OAU/IBAR and the KU, was convened in Arusha, the United Republic of Tanzania, in July 1995.

References

FAO. (2003). *Biosecurity in Food and Agriculture.* Discussion Paper. Committee on Agriculture, 17th Session, Rome.

Knols, B.G.J. & Takken, W., (eds.) (2007). *Emerging pests and vector-borne diseases in Europe.* Wageningen, The Netherlands, Wageningen Academic Publishers

Robertson, A. (1976). *Handbook on animal diseases in the tropics (third edition).* British Veterinary Association. Oxfordshire, UK, Burgess and Sons, LTS.

URT. (2002). *Emergency Animal Disease Surveillance and Control Programme.* Dar Es Salaam, Tanzania, Ministry of Water and Livestock Development.

6

Global Dairy Sector

It has been estimated that in 2005 some 1.4 billion people lived in absolute poverty and that almost 1 billion of them were affected by chronic mal- or undernutrition. Recent food price increases are expected to have pushed many more people – perhaps as many as 100 million – even further into that dire situation. The fight against poverty and hunger is thus a major global concern. Indeed, at the United Nations Millennium Summit of September 2000, world leaders pledged, inter alia, to halve by 2015 the proportion of people living in extreme poverty and hunger.

An estimated 75 percent of the world's poor live in rural areas, and at least 600 million of these people keep livestock to produce food, generate cash income, manage risks and build up assets. With the valuable contribution livestock makes to sustaining livelihoods, especially in rural areas, the development of small-scale livestock enterprises must be seen as a key element of any efforts to eradicate extreme poverty and hunger.

According to data gathered by the International Farm Comparison Network (IFCN), in 2005 around 149 million farm households throughout the world were engaged in milk production. On average, these households keep two milking cows (or buffaloes) yielding about 11 litres/day. Assuming a mean household size of five to six, some 750 to 900 million people (or 12-14 percent of the world population) rely on dairy farming to some extent.

Global Price Trends for Feed and Dairy Products

The key determinants of milk production are world market prices for milk and feed, as illustrated in this section.

World Market Prices For Feed

Prices for corn, as energy feed, and soybean meal, as protein feed, have been used for the purpose of this analysis. In 1981-2006, the world corn price averaged US$109/ton, fluctuating between US$90/ton and US$120/ton. It rose to US$162/ton in 2007 and, in the first six months of 2008, to US$241/ ton or 121 percent above the 1981 to 2006 average and 48 percent over the 2007 price. The time series shows that high prices were recorded in 1996, 2007 and 2008 because of strong demand for food, feed and fuel. The average world market price of soybean meal in 1981-2006 ranged between US$150/ton and US$260/ton, averaging US$212/ton. After the peak in 2004, it stayed close to US$200/ ton until it rose to US$307/ton in 2007 and averaged US$457/ ton in 2008 (January to June). The IFCN feed price indicator - combining corn and soybean meal prices - shows an average of US$140/ton for the period 1981-2006. It rose to US$206/ton in 2007 and to US$305/ton in 2008 (January-June). In 2004, 2007 and 2008 prices were significantly above the historical average compared with relatively low levels in the period 1999 to 2003.

World Market Prices For Dairy Products

The average world market price of butter in 1981-2006 was US$1 580/ton, fluctuating between US$1 000/ton and US$2 000/ton. It shot up to US$2 886/ton in 2007 and, in the first six months of 2008, increased further to US$4 021/ton. Development of the average world market price for SMP showed levels of less than US$1 000/ton between 1981 and 1987; moderate prices of US$1 000 to 2 000/ton, similar to those of 1988 to 2004; and record prices of close to US$2 500/ ton in 2006, US$4 250/ton in 2007 and US$3 750/ton in the first six months of 2008. Prices of SMP in 2006 and 2007 were significantly higher than those of butter but fell below in the first half of 2008. Butter and SMP prices can be converted into prices per kilogram of fresh milk based on assumptions of the processing cost and technical coefficients provided by the Zentrale Market- Fund Preisberichtstelle GmbH (Central Market and Price Reporting Agency, ZMP). Expressed in United States dollars, three periods of 'world market' prices of liquid milk can be distinguished:

Very low – 1981 to 1987: US$8-13/100 kg

Volatile – 1988 to 2006: US$12-26/100 kg

New levels since 2007: more than US$46/100 kg

Expressed in Euro, milk prices stayed at around •15/100 kg, with significant increases in 1989, 2000 and 2001, and with major drops in 1986, 1987 and 1990.

Milk:Feed Price Ratios

From 1981 to 2007, milk prices were more volatile than those of feed. The milk:feed price ratio, which indicates how much feed a dairy farmer can buy with the proceeds of one kilogram of milk, increased steadily from 0.7 kg in 1981 to 2.3 kg in 2007. The price of milk stabilized in the first half of 2008 while that of feed continued to rise and the milk:feed price ratio fell back to 1.5, a level at which low-input milk production systems become more favourable. Milk prices and farm profits were 'high' in 2007 but fell back in 2008, as the milk price development was overtaken by feed price increases, especially in high-input systems. With the new level of milk and feed prices, the milk:feed price ratio will need to be updated.

As far as exchange rates are concerned, this long-term series shows that there was a slight devaluation of the United States dollar against the Euro until 2001 and a stronger one since then. The US dollar was very strong in the periods 1983 to 1985 and 2000 to 2002 but weakened in 2007 and 2008 when it fell below the historic lows of 1992 and 1995.

Milk Production Trends

This section provides both an overview of milk production levels in different parts of the world, and recent trends. The milk production charts are based on an IFCN analysis for 2006-2007 compared with 2002, undertaken in 2008. The analysis was based on milk production surveys (cow and buffalo milk) in 78 countries and on secondary data from organizations such as the Food and Agriculture Organization of the United Nations (FAO). The milk production volumes of all animal species have been standardized to 'energy corrected milk' (ECM, 4.0 percent fat and 3.3 percent protein). The data for milk fat and protein content are based on national statistics or, in the absence of such statistics, on estimates.

Shares in Global Milk Production

World milk production is derived from cows, buffaloes, goats, sheep and camels. As shown in the map in 2007/2006 the major milk production regions are:

- *South Asia:* 23 percent of global production, mainly India and Pakistan.
- *EU-25:* 21 percent, mainly Germany and France.
- *USA:* 12 percent.
- *CIS:* 10 percent, mainly the Russian Federation and Ukraine.
- *Latin America:* 10 percent, mainly Argentina, Brazil, Colombia and Mexico.
- *East and Southeast Asia:* 8 percent, mainly China and Japan.
- *Africa:* 5 percent - the largest milk-producing countries are Egypt, Kenya, South Africa and Sudan.
- *Oceania:* 4 percent.
- *Near and Middle East:* 4 percent, mainly Iran and Turkey.

Trends in Milk Production

During 2002 to 2007, world milk production rose (by 13 percent) to 697 million tons, making for an aggregate increase of 81 million tons or 15 million tons per annum. China, India and Pakistan alone accounted for about two thirds of all volume growth; most of the remaining growth was in Brazil, Egypt, New Zealand, Turkey and the USA. Together, these eight countries accounted for approximately 85 percent of all milk volume growth in 2002 to 2007.

Farmers' Milk Prices and Milk:Feed Price Ratio

For dairy farmers, the most important factor is the producer price for milk. Therefore this section deals with national milk prices and their relation to feed prices in the countries analysed by the IFCN. The analysis covers 2006, the last year before the start of significant increases in world commodity prices.

High Milk:Feed Price Ratios (More Than 2.5)

Highly favourable milk: feed price ratios of more than 2.5 are found in Canada, Egypt, Greece, Kazakhstan, Mongolia, Saudi Arabia, Sudan and the USA. In most cases, the cause of a high milk:feed price ratio is a very high milk price (up to US$30/100 kg) while in a few cases it is caused by feed prices significantly below the world market level (such as in Kazakhstan).

Milk Prices Per Country 2006

Milk prices per country range from US$15 to 74/100 kg ECM and can be grouped into five categories:

- *US$20:* New Zealand, Argentina, Uruguay, Paraguay, Uganda, Belarus, Ukraine, Pakistan and Indonesia.
- *US$20 to 25:* Australia, Uzbekistan, Nigeria, Brazil, Chile, Bolivia, Peru, India and Lithuania.
- *US$25 to 30:* China, Viet Nam, Poland, Bulgaria, Romania, Turkey, Russia, Kazakhstan, Kenya, South Africa, Colombia, Ecuador and a number of Central American countries.
- *US$30 to 40:* USA, Mexico, Venezuela, most EU countries, Hungary, the Czech Republic, Estonia, Slovenia, Slovakia, Israel, Iran, Mongolia, Morocco, Algeria, Tunisia, Ethiopia, Cameroon, Thailand, Myanmar, Malaysia and The Philippines.
- *US$40:* Canada, Iceland, Norway, Finland, Switzerland, Italy, Greece, Egypt, Sudan, Saudi Arabia, Mozambique, Taiwan, South-Korea and Japan.

Method Milk:Feed Price Ratio

The milk:feed price ratio as defined by IFCN as the milk price divided by the price of purchased feed. In simplified form, it indicates how much feed (in kilograms of concentrate) it is possible for a farmer to buy with the sale proceeds from one kilogram of milk. The higher the ratio, the more economical it is to use concentrates to feed the dairy cows. Currently IFCN regards the ratio as favourable for the use of concentrates when it is higher than 1.5, which is when high-input high-yield dairy systems become profitable.

Intermediate Milk:Feed Price Ratios (1.5-2.5)

Most countries of Europe and the Commonwealth of Independent States (CIS) fall into this category, as well as Argentina, Brazil, Ecuador, Ethiopia, India, Japan, Republic of Korea, Mexico, Morocco and Viet Nam.

Low Milk: Feed Price Ratios (Less Than 1.5)

Very unfavourable milk:feed price ratios (of less than 1.0) have been observed in Cameroon, Guatemala, Indonesia, Nigeria and Uganda, whereas they are slightly better (1.0 to 1.5) in Australia, China, Chile, Ireland,

Myanmar, New Zealand, Norway, Pakistan, Peru, South Africa, Switzerland, Thailand, Turkey and Uruguay. In most cases, the causes of unfavourable milk:feed price ratios are low milk prices (less than US$20/100 kg). In a few cases, they are caused by feed prices significantly above the world market level, such as in Switzerland and Norway.

From the milk:feed price ratio, it is possible to obtain an indication of which types of dairy farming systems fit best into a given country or region. For instance, a high milk feed:price ratio indicates that it may be profitably to intensify a farming system. Once the milk:feed price ratio starts to fall – driven either by falling milk prices or increasing feed prices – 'extensifcation' of the system might be preferable.

Dairy Farm Numbers World Wide

This section describes the structure of the dairy sector in selected countries, in terms of farm numbers and average dairy herd size.

Dairy Farm Numbers

In 2005, there were some 115 million dairy farms in the 73 countries for which the IFCN has detailed information. Based on this number IFCN estimated a total number of dairy farms for 2005 of 149 million considering all countries. Assuming that the average farm household comprises fve to six persons, about 750 to 895 million people, or 12 to 14 percent of the world population, directly depend to some extent on dairy farming.

The number of dairy farms is highest in India and Pakistan (75 and 14 million, respectively), followed by Brazil, China, Ethiopia, Iran, Romania, Russia, Turkey, Ukraine and Uzbekistan with 1.0 to 2.5 million dairy farms each. Farm numbers in the EU-15 countries (533 851) and the USA (78 300) seem rather low in comparison.

Average Dairy Herd Size

The development of dairy farm numbers shows two trends. In Argentina, Australia, Brazil, Europe, Japan, New Zealand, South Africa and the USA, numbers dropped by 2 to 10 percent per annum between 2000 and 2005 compared with annual increases of 0.5 to 10 percent in most developing countries.

The development of dairy farm numbers shows two trends. In Argentina, Australia, Brazil, Europe, Japan, New Zealand, South Africa and the USA, numbers dropped by 2-10 percent per annum between 2000 and 2005 compared with annual increases of 0.5-10 percent in most developing countries.

Dairy Herd Sizes

IFCN estimates that, globally, the average dairy herd size is 2.4 cows. In most countries, especially in Africa, Asia, Eastern Europe and parts of Latin America, the vast majority of dairy farms comprise less than ten cows, and only 15 countries have an average dairy herd size of more than 50 cows. The six countries with average dairy herds comprising more than 100 cows are: Argentina, Australia, Czech Republic, New Zealand, South Africa and the USA. In most countries, average dairy herd sizes (0 to 5 additional cows per farm) did not change significantly in 2000-2005. The greatest increases during that period were observed in New Zealand (+79), Australia (+42), USA (+28), Denmark (+25), South Africa (+19), Israel (+16) and the Netherlands (+10).

Pattern of Dairy Trade and Milk Processing

This section describes the pattern of world dairy trade, the purpose being to identify the major dairy exporters and importers and illustrate the degree of self sufficiency and milk processing structure by country. The analysis is based on that of the IFCN undertaken in 2006 covering the period 1990 to 2004. It should be mentioned, however, that the core competence of the IFCN relates more to milk production rather than to trade and milk consumption.

Milk Self-SUFFICIENCY, Surplus and DEFICIT

Few countries are self-sufficient in milk, which means they import more dairy products than they export. Very low self-sufficiency rates in milk (less than 25 percent) were observed in Bahrain, Democratic Republic of the Congo, Côte d'Ivoire, Gabon, Gambia, Ghana, Jamaica, Kuwait, Liberia, Malaysia, Papua New Guinea, Philippines, United Arab Emirates and Viet Nam.

Top Ten Net Exporting/Importing Countries

The following table shows the largest net milk exporters/ importers in 2003-2004. It should be noted that the list is based on net trade figures, that is, the balance of exports of milk after subtracting the quantities imported converted to ME.

Table 1: Top Ten Net Exporting/Importing Countries

	Net Exporters	*Net Importers*
1	New Zealand	China
2	EU-15	Mexico
3	Australia	Japan
4	EU-10 New members	Algeria
5	USA	Russian Federation
6	Argentina	Philippines
7	Ukraine	Saudi Arabia
8	Belarus	Indonesia
9	Uruguay	Nigeria
10	Switzerland	Viet Nam

Share of Milk Processed in Tradable Dairy Products

Tradable dairy products comprise condensed milk, cheese, dry milk products, butter/ghee, which, due to processing are far less perishable (and bulky) than liquid milk. A high share of tradable dairy products in relation to national milk production indicates that a considerable amount of milk passes through the formal sector, but also that the national dairy industry is exposed to competition from other countries in a liberal agricultural trade environment. Globally, countries can be divided into three groups with respect to the shares of milk processed into tradable products:

High Shares (More Than 50 Percent):

Australia, Belgium, Czech Republic, Denmark, France, Germany, Ireland, Netherlands and New Zealand convert more than 50 percent of their milk production into tradable dairy products.

Moderate Shares (30-50 Percent):

Results of around 30 to 50 percent were observed for Argentina, Chile, Estonia, Italy, Finland, Hungary, Iceland, Japan, Republic of Korea, Lithuania, North America, Peru, Poland, Sweden, Switzerland and Venezuela.

Low Shares (Less Than 30 Percent):

In developing countries the share of milk processed into tradable dairy products is rather low (0 to 20 percent), as seen for instance in Africa, Asia and countries of Latin America. Low shares have been also observed for Spain, Ukraine and Russia.

Share Of Milk Production Traded

Based on the analysis 2004 about 7.1 percent or world milk production is traded internationally (Intra-EU trade excluded). With respect to milk delivered to milk processors, we estimate the share traded internationally to be in the order of 24 percent.

MILK CONSUMPTION AND ITS DRIVERS

Milk demand is driven by two factors: per capita milk consumption and population. The aim of this section is to give a global overview of both indicators via world maps, with a description of country-specific differences.

Method – *Per Capita* Milk Consumption

The method used to calculate per capita consumption is described in the IFCN Dairy Report 2004, which is based on 'milk equivalents' (MEs) so as to account for the consumption of milk in its different forms, such as yoghurt or cheese, in addition to liquid milk. The per capita consumption was calculated as follows: milk production (in ME) minus exports (in ME) plus imports (in ME) plus/minus changes in stocks (in ME) divided by human population. The 'total solids' method was used to convert dairy products into ME.

Population STATUS 2004 and Trends

About 60 percent of the world population live in South, East and South-East Asia, with China and India alone accounting for about 38 percent. Another 14 percent is to be found in Africa. In all these countries (except India, Pakistan and some African countries), milk consumption is generally below 30 kg of milk (ME) per capita. Western Europe and North America account for 11 percent of the world population with an average per capita consumption of approximately 300 kg of milk (ME) per year.

Examples of Milk Demand Growth

Some simple examples illustrate how milk demand can develop: once milk consumption in China (2004 = 22 kg of milk (ME) per capita) increases to the level of Japan (78 kg of milk (ME) per capita) it will require about 72 million tons of milk, which is almost equal to the production volume of the

USA. Once milk consumption all over India increases from 93 kg milk per capita to the level typical of the richer states of Punjab and Haryana (IFCN estimate 200 to 250 kg milk per capita), this will call for an additional 17 million tons of milk -which is more than the EU-25 was producing in 2006.

Per Capita **Milk Consumption Per Country**

As a general rule milk consumption is high in developed countries and low in the developing ones, and appears to be particularly low in tropical and subtropical climates. Based on country-specific estimates of per capita milk consumption, the following three categories have been defined:

High, More Than 150 kg Per Capita/year:

Argentina, most CIS countries, Costa Rica, Ecuador, Europe, Honduras, Israel, Lebanon, North America, Oceania, Turkey, Uruguay and others such as Pakistan and Sudan.

Medium, 30-150 kg Per Capita/year:

India, Japan, Republic of Korea, North and Southern Africa, most countries of the Middle East and Latin America.

Low, Less Than 30 kg Per Capita/year:

China, Ethiopia, Yemen and most countries of Central Africa and East and Southeast Asia

The Two Drivers of Milk Demand

In past years, milk consumption has risen by 10 to 20 million tons per year, one driver being human population growth. A global population growth rate of 1.2 to 1.3 percent per year means 75 to 80 million more people each year. Using the world average per capita milk consumption, this would mean that population growth accounts for an increase in milk consumption of 7 to 9 million tons per year. The second driver of milk consumption is increasing per capita consumption. However, this driver in turn depends largely on per capita income developments, especially in developing countries.

Milk Production and Dairy Sector Profiles

This section contains a country-by-country analysis of the status of, and developments in, national dairy sectors and provides the wider perspective for the detailed farm-level analysis . Because the availability and quality of data in most developing countries is problematic, the time frame chosen for this analysis, 1996 to 2005, relates to information contained in the IFCN Dairy Reports, 2006 and 2007. The country profiles provide an overview of a number of indicators illustrating the trends and drivers for milk supply and demand, and the dairy chain. The intention is to give each country's dairy sector a 'face'. In all cases, it has been attempted to make the indicators comparable between the countries.

For the purpose of this analysis, ten developing countries were chosen as well as three developed dairy countries to put the developing countries analysed into a global context. The developing countries are Bangladesh, Cameroon, the People's Republic of China (henceforth China), India, Morocco, Pakistan, Peru, Thailand, Uganda and Viet Nam. Comparable data were available because the IFCN is well established there.

India

With an annual production of 108 million tons of ECM, 65 percent of which is produced by buffaloes, and a national herd of 113 million head of cattle/buffaloes, India is the world's largest milk-producing country. Some 75 million dairy farming households, with an average of 1.5 adult female cows or buffaloes per farm, are engaged in the sector each producing about 4 litres of milk per farm/day. During the period under review, production rose by 3 to 4 percent per annum or approximately 4 million tons, thanks to higher milk yields and more cows and buffaloes.

The predominant dairy production systems may be classified as low-input/low-yield systems (956 litres/cow/year). Feeding is based mainly on crop residues such as straw and green fodder, supplemented by small quantities of low-cost compound feed. Milking is done by hand and the milk transported to village collection centres or collected by local milkmen. About 45 percent of the milk is used by the farming households and only 15 to 20 percent is delivered to formal milk processors. Annual per capita milk consumption increased by 1.5 to 2.4 percent per annum from 1990, reaching 98 kg in 2005.

Previously, rising demand for milk was mainly driven by population growth whereas increases in per capita consumption have now become an additional driver. India has always been 100 percent self-sufficient in milk, with total imports/exports of only 0.3 million tons per annum; it may thus be considered as almost unconnected with the world dairy market.

Pakistan

With a production of 34.4 million tons of ECM, Pakistan was the world's third largest producer of milk in 2005, with buffaloes accounting for 75 percent of production. Milk is produced by approximately 15 million dairy farming households with an average of 1.8 adult cows or buffaloes per farm producing approximately 6.4 litres of milk per farm/day. Between 2000 and 2005, production grew by 2.9 percent per annum, thanks more to increased numbers of milking animals than to higher milk yields.

Dairy production systems in Pakistan are similar to those in India. Most (50 percent) of the milk is consumed by the farming households or sold on the informal market (40 percent); less than 10 percent is delivered to formal milk processors.

By 2005, yearly milk consumption in Pakistan had reached 230 kg per capita, significantly higher than in India. Increased demand for milk was mainly driven by population growth (from 2.0 to 2.2 percent per annum). Like India, Pakistan has always been completely self-sufficient in milk, with imports/ exports of only 0.22 million tons per annum.

Bangladesh

Dairy production systems in Bangladesh are similar to those in India and Pakistan. However, milk production and yields (2.8 million tons ECM from cows and buffaloes, and 711 kg of ECM per cow/per day, respectively) are significantly lower than in India and Pakistan.

Most of the milk is consumed by farming households or sold on the informal market, and less than 20 percent is delivered to formal milk processors. In 2005, per capita milk to 10 litres in 2005. Viet Nam is currently 25 percent self-sufficient in milk, and imports about 0.6 to 0.8 million tons per year. consumption stood at only 32 kg/year. Bangladesh is 85 percent self-sufficient in milk and imports 0.4 million tons per annum.

Thailand

In 2005, Thailand produced 0.8 million tons of ECM, less than 1 percent of that produced by India. Nevertheless, with an annual increase of 8.4 percent, production has increased rapidly since 2000, mainly thanks to greater numbers of cows.

With an average of 20 cows per farm, Thailand's dairy herds are significantly larger than those in Bangladesh, India and Pakistan. Moreover, the country's dairy farming systems are more intensive than in other parts of South Asia owing to its development policy and high milk prices (about 30 to 40 percent above those in India). Dairy production relies mostly on Holstein cows that have higher milk yields than the buffaloes or local cows used in Bangladesh, India and Pakistan. Milking is mainly done by machine and about 95 percent of the milk is delivered to formal milk processors.

In 2005, yearly milk consumption stood at 21 kg per capita. Thanks to its substantially increased production, the country's milk self-sufficiency increased from 33 percent in 1996 to 47 percent in 2005. Nevertheless, Thailand's annual milk deficit stands at approximately 1 million tons.

Viet Nam

With a production level of 0.23 million tons of ECM in 2005, Viet Nam is the smallest milk producer of the Asian countries covered by the analysis. However, during the period under review, milk production grew by more than 20 percent per annum, mainly driven by increasing milk yields that had reached 1.73 tons per cow/year by 2005.

On average, dairy farms in Viet Nam have 6.9 cows producing 32 litres of milk per farm/day. Production is mainly based on imported dairy cattle or crossbreds with local cattle. As in Thailand, about 95 percent of Viet Nam's milk is delivered to formal milk processors. Per capita milk consumption increased from 4 litres in 1996

China

In 2005, China was the world's fifth largest producer of milk, accounting for 24.5 million tons of ECM from cows and (to a lesser degree) buffaloes. Based on yearly increases of 27.2 percent in the production of cow's milk over the period 2000 to 2005, China should rapidly become the world's third largest milk producer. Moreover, as most of the milk is sent to formal processors, China will soon rank second in terms of milk processing volumes. Production growth has been driven mainly by increased numbers of cows rather than increased milk yields.

With an average of 3.7 tons per cow/annum, China's milk yields are the highest of all the Asian countries covered by the analysis. While the average herd size stands at 6.7 cows, Chinese dairy farms fall into two categories: small farms with 1 to 40 cows; and large farms with more than 200 cows. The small farms usually deliver their milk to a local collection point, take their cows to village milking centres or belong to a 'dairy garden' for which investors have provided the basic dairy infrastructure. The larger farms are either operated by the state (mainly in the southeast) or by private investors with close ties to the major dairy companies. As most dairy farms in China have insufficient land, farmers are obliged to purchase compound feed and roughage, the latter mainly in the form of corn silage.

Annual per capita milk consumption increased from 8 litres in 2000 to 22 litres in 2005 and to an estimated 28 litres in 2007. Of all the milk consumed in China, 86 percent is produced within the country.

Uganda

In 2005, Uganda's 0.8 million dairy farmers, with an average of 2 cows/ farm yielding 3.6 litres of milk per farm/day, produced 1.4 million tons of ECM. Annual milk production has risen by 13.1 percent since 2000, mainly thanks to increased milk yields (from 510 kg/cow/year in 2000 to 800 kg/ cow/year in 2005). Milk supply in Uganda is very seasonal, peaking in April with 125 percent of the yearly average and at its lowest in June/ July with only 65 percent of the yearly average.

Uganda's dairy farming systems may be classified as low-input/low-yield. Feeding is based mainly on grazing supplemented by small quantities of low-cost compound feed. Milking is done by hand and the milk transported to milk collection centres in villages or collected by local milkmen. About 30 percent is consumed on-farm. In 2005, annual per capita milk consumption stood at 50 kg, increasing by 4 to 6 percent per annum. As yearly population growth is in excess of 3 percent, it follows that national milk demand is increasing by 8 to 10 percent per annum. Uganda is currently self-sufficient in milk and neither imports nor exports significant volumes. Only 2 percent is delivered to milk formal processors.

Cameroon

With 0.13 million tons of ECM produced in 1996-2005 by approximately 4 000 dairy farmers, milk production and yields in Cameroon are lower than in Uganda. According to official statistics, production in Cameroon remained stable between 1996 and 2005, contrary to claims of increases on the part of local dairy experts.

As a general rule, milk production in Cameroon is a secondary activity of larger cattle herds that are kept for beef production. Feeding is mainly based on grazing and no use is made of compound feed. Milking is done by hand, and only 2 percent of the milk is delivered to formal milk processors.

In 2005, yearly per capita milk consumption stood at 14 kg but, according to official statistics, is declining. In the same year, Cameron imported about 23 percent of its milk needs.

Morocco

The country's dairy sector is very similar to that of Uganda. In the period under review, some 1.4 million tons of milk were produced by about 0.8 million dairy farmers with an average of 2 cows/farm. Milk production estimated to be growing at about 4.2 percent per annum.

Milk production in Morocco is usually a side activity of crop farmers cultivating around 2 ha of land. The feeding system is similar to that in India/ Pakistan and is mainly based on compound feed and green fodder. Milking is mostly done by hand and, in 2005, about 63 percent of the milk was delivered to formal milk processors. In 2005, per capita milk consumption stood at 62 kg. Morocco is a net importer of dairy products (0.4 million tons ME), and is 80 percent self-sufficient in milk.

Peru

In 2005, Peru produced 1.27 million tons of ECM on 108 000 dairy farms, with an average of 6.4 dairy cows/farm producing about 32 litres of milk per farm/day. This shows a yearly growth of 4.5 percent, of which the main determinant was a 6.5 percent increase in the number of cows in 2000 to 2005. Over the same period, however, yearly milk yields per cow decreased from 2 000 kg to 1 850 kg.

Dairy farming systems may be classified as low-input/low-yield. Feeding is based mainly on grazing supplemented by small quantities of low-cost compound feed. Some milk is produced on intensive dairy farms, mainly in the coastal region. Milking is done by hand and the milk transported to milk collection centres in villages or collected by local milkmen; about 94 percent is delivered to formal milk processors.

In 2005, annual per capita milk consumption stood at 51 kg. Between 2000 and 2005, increased demand for milk was mainly driven by population growth (1.5 percent/year). Peru is approximately 93 percent self-sufficient in milk.

Germany

Germany was the world's fourth largest producer of milk in 2005, accounting for 29.5 million tons of ECM, and the second largest milk processor (behind the USA). Milk is produced by 110 000 dairy farmers with average herds of 37.6 cows producing 732 kg of milk/day (19.5 kg/cow). National milk production has been stable since 1990 because of the milk quota system. Yields increased by 2 percent per annum in 2000 to 2005, although the number of dairy cows decreased by 2 percent per annum over the same period.

The country's dairy production systems may be classified as high-input/high-output (7 100 litres per cow/year). Feeding is based mainly on grass/corn silage and compound feed. Milking is done by machine, after which the milk is stored on-farm in cooling tanks and collected by local milk processors every two days. About 95 percent is delivered to milk processors; the remainder is either used on the farms (for home consumption or for feeding calves) or is sold directly to consumers.

Having remained stable since 1996, the country's annual per capita consumption stood at 309 kg of ECM in 2005. As a member of the EU, Germany exports about 40 percent of its milk and imports some 30 percent of its consumption needs. The country is 116 to 127 percent self-sufficient in milk, which translates into a surplus of 4 to 6 million tons per annum.

United States of America

The USA produces 76 million tons of ECM/year, generated by 78 000 dairy farms with average dairy herds of 115 cows producing 2 643 litres/day (or 23 litres/cow). Since 1975, national milk production has grown steadily by 1.1 percent per annum, driven by yield increases of 1.5 percent and a 0.3 percent reduction in the number of dairy cows.

As high-input/high-output (8 400 litres per cow/year). As in Germany, feeding is based mainly on grass/corn silage and compound feed. The cows are milked by machine, mainly in milking parlours, and the milk is stored on-farm in cooling tanks before being sent to formal processors. About 99 percent is delivered to processors.

Since 2000, annual per capita milk consumption has remained stable at around 250 kg of ECM. In 2005, the USA exported about 3.4 percent of its milk and imported 2.8 percent of its internal demand. Self-sufficiency stood at around 104 percent in 2000 to 2005, translating into an annual milk surplus of 3 to 5 million tons.

New Zealand

In 2005, New Zealand produced 15.8 million tons of ECM, corresponding to about 20 percent of that in the USA. This was produced by 12 300 dairy farmers with average dairy herds of 315 cows yielding 3 526 kg/ day (or 11.2 kg/cow). Production increased by 4.6 percent per annum in 2000 to 2005, mainly driven by increased numbers of cows, The country's dairy production systems may be defined as intermediate-input/intermediate-output (3 868 litres per cow/ year). Feeding is based mainly on grazing. Milk production is therefore seasonal, peaking in November (180 percent of the annual average) and at its lowest in June and July (5 to 10 percent). Milking usually takes place in swing-over parlours or rotary milking systems, after which the milk is stored in cooling tanks on-farm and subsequently collected by local milk processors. Almost 100 percent of the milk is delivered to formal milk processors.

New Zealand exports about 95 percent of its milk production and, with an export volume of about 15 million tons, it is the world's largest exporter of the commodity.

International Competitiveness of Dairy Farms

This section compares the international competitiveness of 'typical' dairy farming systems in selected countries, with special focus on the cost of milk production and main determinants thereof. It also seeks to determine whether small-scale dairy farms in developing countries are in a position to compete with large-scale dairy farming systems in industrialized countries.

Thirty dairy farm types from 13 countries have been compared using the standard IFCN methodology. Farms representative of various dairy farming systems were selected in Bangladesh, Cameroon, China, India, Morocco, Pakistan, Peru, Thailand, Uganda and Viet Nam and subjected to detailed analysis. For the industrialized countries, similar analyses were conducted for farms in Germany, New Zealand and the USA. In general, the first farm 'type' represents the current 'average' farm, while the second represents either a larger farm type or a different, but relevant, dairy production system. While production data usually refer to the calendar year 2005, data for the farms in Africa and China were obtained in 2006 when they entered the IFCN data collection system. For the farms in India (Orissa and Karnataka), Thailand and Viet Nam, the data were obtained in 2004. In order to make use of available information, 2005 exchange rates were applied to all financial farm data.

Comparison of Dairy Returns

Dairy farm returns derive from milk and/or non-milk items (sales of cattle and manure, government payments, etc.). Milk returns account for 55 to 95 percent of the returns of all farm types analysed, and range from US$12 to US$36/100 kg of ECM. These returns are mainly determined by three categories of farmgate milk prices, which were:

- *Less than US$20/100 kg:* Observed in Pakistan and Uganda.
- *US$20-30/100 kg:* Although most farms receive prices in this range, those in India, New Zealand Viet Nam are at the lower end of the range whereas prices in Bangladesh and Thailand are at the higher end.
- *More than US$30/100 kg:* Farmers in Cameroon, Germany, Morocco and the USA all obtained similar milk prices of about US$36/100 kg of ECM.

Non-milk items account for US$2 to 38/100 kg ECM of the returns of the dairy farm types analysed. The main determinants of non-milk returns were the cattle/beef price levels; culling rates and, related to that, strategies for selling calves and surplus heifers; yields per lactation; the level of government payments coupled to milk production; and use of manure on the farm. Based on these factors, non-milk returns were very low for the farms in India and very high in Germany and Morocco.

Comparison of The Cost of Milk Production

Lowest milk production costs (less than US$12/100 kg) were observed for both farm types in Uganda and for the larger farm type in Cameroon. Production costs on all farm types in China, New Zealand, Peru, Thailand and the USA were slightly higher (US$22 to 30/100 kg), but were highest in Germany (over US$35/100 kg).

Given major differences in agricultural wage rates between developed and developing countries, it might be assumed that developing country farms have a labour cost advantage. It is, however, interesting to note that this was found not to be the case when comparing labour costs per litre of milk, mainly because as a general rule regions with higher salaries also have a significantly higher level of labour productivity. Expressed in terms of per litre of milk, the labour costs of a nine-cow dairy farm in Punjab, India, are similar to those of a 350-cow farm in the USA. The key cost advantage of smallholder dairy farming is the use of low-cost feed and the overall 'low-tech' approach to milk production. Cows fed on crop residues, such as straw, are lower-cost producers of milk than high-yielding, grain-fed dairy cows.

Economies of Scale

In general, comparison of farm types within and across countries and regions supports the hypothesis that economies of scale exist in milk production. It was observed that in Thailand and Uganda lower milk production costs were incurred only by the smaller farm types, mainly because of their significantly higher non-milk returns per 100 kg of milk.

To obtain a clearer understanding of their comparative position in terms of international and local competitiveness, a simple average of key indicators of competitiveness was calculated for farmers in both developing and developed countries. Although the method is very crude and there are very large variations within each group, the comparison is most informative.

Average milk production costs in the three developed countries covered by this study (Germany, New Zealand and the USA) stand at US$31.4/100 kg, or 56 percent above the average production cost of US$20.2/100 kg calculated for the ten developing countries (Bangladesh, Cameroon, China, India, Morocco, Pakistan, Peru, Thailand, Uganda and Viet Nam). The average price of milk in the three aforementioned developed countries (US$31.2/100 kg) is only 30 percent higher than that of the developing countries (US$24.0/100 kg). Thus, the overall profitability of milk production appears to be higher in developing countries than in the developed ones, which may be one of the reasons why developing countries are increasing their shares in global dairy production.

If dairy farming is to be sustainable, it is essential that it should be able to compete for labour on the local labour market. The indicator 'return to labour' quantifies the 'value-added' per hour of labour put into dairy farming. If this return is higher than the average local wage rate, then the farming system can afford to pay competitive wages and should be sustainable from the labour standpoint. The average return to labour observed in the developing countries covered by this study is US$0.45/hour, which is 45 percent higher than the average local wage of US$0.31/hour. In the three developed countries, the average return to labour is US$16.30/hour, which is still 22 percent above the average estimated wage of US$13.30/hour. These figures indicate that dairy farming can compete on the local labour market in both groups of countries. However, despite these favourable returns to labour from dairy farming, the figures also show that milk production can quickly lose its competitive advantage if local wages rise faster than labour productivity.

Overview of The Whole Farm

Most of the farm types included in the assessment were specialized dairy farms obtaining more than 60 percent of their returns from dairy activities. Non-specialized dairy farms (with less than 60 percent of their returns from dairying) were observed in Bangladesh (Sirajgani), India (Orissa and Punjab), Pakistan (Layyah) and Uganda (Kayunga), mainly because dairying is a supplementary occupation in mixed farming systems.

Many dairy farms also have crop enterprises that produce mainly cash crops. Other typical non-dairy activities include sheep- or goat-raising (PK-1, BD-2, UG-13, PE-6): poultry production (IN-4KA, PK, BD-2, TH-14, UG-3); pig-raising (UG); -farming (BD-10); contract labour/custom work (DE-30S, US-80WI); and forestry (DE-30S). Of-farm activities are very common on small farms in Bangladesh, India, Morocco, Pakistan, Thailand, Viet Nam and Uganda. The DE-30S farm obtained additional income from renting out houses.

Profit Margins

Proft margins range from -36 percent to +70 percent. As a general rule, margins are high (more than 25 percent) on small family farms because they are run without hired labour and have limited liabilities.

Farm Income, Profits and Returns to Labour

Farm Income

The 'farm income' indicator describes a farm's income based on the proft and loss (P&L) account. On family farms, this income provides the basis for covering family livelihoods and asset or capital growth. Salaries are deducted for farms that have hired labour. Farm income ranges from US$3/100 kg of milk to US$38/100 kg of milk. Smaller farms tend to have higher farm incomes per 100 kg of milk because of the higher proportion of family labour, which is not deducted. In Cameroon and Viet Nam, however, larger farms with hired labour generate higher farm incomes than the smaller farms, mainly because the former obtain higher output prices (Viet Nam) or have low wage costs for hired labour (several larger farms in Cameroon share a herdsman).

Entrepreneur Profit

The 'entrepreneur proft' indicator shows whether farms cover their full economic costs. If the indicator is positive, all costs shown in the P&L account can be covered and family-owned production factors (labour, land, capital and quota) can be paid at market price (opportunity cost). In this case, farming systems are financially sustainable. Entrepreneur profits range from -US$15/100 kg of milk to +US$34/100 kg of milk; some 73 percent of the farms make a positive entrepreneur proft. Negative profits (losses) are found on the smaller farm types in Morocco, New Zealand, Pakistan and Peru, and on both farm types in China and Germany. In nearly all countries except Thailand and Uganda, the larger farm types earn greater profits per 100 kg of milk than the smaller ones. As mentioned earlier, the main reasons why the smaller farm types in Thailand and Uganda make higher profits predominantly relate to their higher non-milk returns due to high prices for cattle (and beef).

Returns To Labour

The 'returns to labour' indicator shows the 'value' created (per hour) by a farm labourer and indicates the potential wage that can be paid. A comparison of returns to labour with the wage rate in the region concerned shows whether a farm can compete on the local labour market and cover its full economic costs. Farms unable to compete with wage levels in their regions may stay in business until a generation change takes place or as long as farmers are satisfied with the 'wages' they obtain for their work. For reasons of simplification, three general levels can be defined:

- *Less than US$1/hour:* All farm types in the developing countries, except the larger farm types in Cameroon and Peru, have returns to labour of less than US$1.00/hour to put back into the dairy farm enterprise. This group represents 73 percent of the farms analysed here.
- *Between US$1 to 10/hour:* Farms CM-35, PE-15, DE-30S and NZ-282.

- *More than US$10/hour*: The larger farm types in Germany and New Zealand and both farm types in the USA.

Competitiveness on The Labour Market

Farms are competitive on the local labour market if their returns to labour input exceed wage levels in their regions. In Asia, the large farm types and both small and large farm types in Thailand and Viet Nam have competitive returns to labour but neither of the Chinese farms can compete on the local labour market, mainly due to the relatively high wage levels in the region. Diversity of local wage levels is large in Africa, as are returns to labour. Returns to labour in dairy farming are locally competitive only in Cameroon due to the grazing practices that keep labour costs low and because of the relatively high amount of beef/livestock sales that make a huge impact on dairy profitability. In Peru, the larger peri-urban farm type generates competitive returns to labour in relation to the local wage rate.

In the industrialized countries, the larger farm types generate higher returns to labour than the smaller ones and are thus more competitive on the local labour market. The most profitable farms were found in the USA, followed by the large farm type in New Zealand. Both farm types in Germany would need to undergo major restructuring if they are to generate locally competitive returns to labour. The overall difference between the highest- and lowest-performing dairy farms in terms of returns to labour input, is in the order of US$32.6/ hour (US-350WI with US$33/hour and IN-9PU with US$0.4/ hour).

Producer Milk Prices and Non-milk Returns

Milk Prices

Milk prices per 100 kg of ECM range from US$12 to US$36 and may be grouped into the following categories:

- *Less than US$20/100 kg*: Observed in Pakistan and Uganda for farms far from urban centres and in areas with poor infrastructure. The lowest price was observed in Ugandan farms at US$12/100 kg of ECM
- *Between US$20 to 30/100 kg:* The majority of the farms studied receive milk prices that fall into this category. Farmers in India, New Zealand and Viet Nam are at the lower end of the scale, whereas farmers in

Bangladesh and Thailand obtain prices at the higher end. In Peru, one farm is located in a peri-urban area and the other in a rural area, which explains the very significant price differences within the country.

- *More than US$30/100 kg:* Farmers in Cameroon, Germany, Morocco and the USA all obtain milk prices of around US$35/100 kg of ECM.

A dual milk price structure appears to exist at the global level, with most farms obtaining prices around the world market level. (The 'world' market price for milk can be calculated by taking the average world market price for butter and SMP, multiplying it by technical coefficients and subtracting processing costs. In 2006, this price stood at around US$26.5/100 kg.) The Moroccan and Cameroonian farms analysed receive prices above those of the world market, similar to those received by their German and US American counterparts. In Morocco, the key driver is the tariff protection, which is similar to that in the EU. Cameroon appears to represent as a special supply/demand situation with regard to fresh milk.

Non-milk Returns

In addition to the income received from milk sales, dairy farms also obtain revenue by means of direct payments, sales of livestock, and sales of manure as fertilizer and household fuel for domestic use (South Asia). These non-milk returns range from US$1.6 to US$35/100 kg of ECM and contribute from 6 to 51 percent of total farm returns. High shares were observed in China (CN-3), Morocco (both farm types) and Uganda (UG-3), mainly as a result of high prices for beef and heifers and very low milk yields (Uganda).

Direct Farm Support

Among the farms studied, direct payments are found only in Germany and the USA, where the level of support is US$6.5 to 11.0 and US$0.7/100 kg of ECM, respectively. No direct support payments are received by farmers in Africa, Asia, Latin America and Oceania.

- *Germany:* Farms in southern Germany were found to receive higher direct payments. These farms are located in less favourable areas and receive payments for extensive use of grassland or for farming in hilly areas.
- *USA:* Dairy farmers cultivating feed grains or soybean receive various subsidies in the form of fixed and deficiency payments. For maize, these payments represented about 40 percent of the market price in 2005.

Support for Investments in Dairy Farming

In many countries, the government and/or non-governmental institutions provide support to the dairy sector by means of investment aid or subsidized interest rates. Although many specific support programmes for dairy development have been set up in both the industrialized and developing countries, there also appears to be a large number of 'not-so-evident' programmes in most countries studied. Thus, direct payments constitute only a small part of measures in support of dairy farming.

Labour Costs

Labour Costs/100 kg Milk

Labour is one of the main costs incurred in dairy farming. To obtain a better picture of the structure of labour costs, the opportunity costs and wages are shown separately in the graph. Total labour costs range from US$2 to US$25/100 kg of milk. Different levels can be discerned:

- *High (more than US$10/100 kg milk):* The farms in China, Germany, and the smaller farm types in Bangladesh, Orissa (India), Morocco and Pakistan
- *Intermediate (US$5 to 10/100 kg milk):* Both farm types in Uganda, Cameroon and USA; the small farm types in Punjab (India) and Viet Nam; and the larger farm types in Bangladesh, China and Morocco.
- *Low (less than US$5/100 kg milk):* Both (small and large) farm types in Karnataka (India), New Zealand and Thailand, and the large farms in Orissa and Punjab (India), Pakistan and Viet Nam.

Labour cost differences may be explained by differences in wages and/or differences in labour productivity. For example, labour costs per 100 kg of milk on the large New Zealand farm type (NZ-1042, US$4/100 kg) amount to 30 percent of those on the small German dairy farm type (DE-80N, US$13/100 kg). The New Zealand farm pays only 60 percent of the German wages (US$10.81/hour vs. US$18.19/hour) and produces 2.3 times the amount of milk per hour of farm labour (321 kg/ hour vs. 140kg/hour).

Labour Productivity

Apart from wage levels, labour productivity also determines the level of a farm's competitiveness and may be influenced by the farmer and the production system adopted. In line with regional wages for farm labour, three labour productivity levels, expressed in kilograms of milk per hour worked, were observed:

- *High (more than 50 kg/hour)*: Found on all farms in industrialized countries (Germany, New Zealand and the USA). On these farms, labour is used very efficiently thanks to labour-saving mechanization and economies of scale. However, even within dairy farms in industrialized countries, large differences in labour productivity exist, ranging from 70 kg/hour on the larger German farm type to as much as 321 kg/hour on the large New Zealand farm.
- *Low (10 to 50 kg/hour):* This small group consists of both farms in Thailand and the smaller farm type in Germany. Variations within the group are determined mainly by milk yields, wage levels and management approach.
- *Very Low (less than 10 kg/hour):* Key factors explaining low labour productivity in Africa, Asia (except for Thailand) and Peru are very low wages and limited access to capital, which result in labour-intensive and capital-extensive production systems with low investments in labour productivity-enhancing equipment.

Land Costs

Land costs contribute between 2 percent and 30 percent of the total costs of dairy farms that rent or own land, and range from US$0.05 to US$8.1/100 kg of milk. In many instances land costs are extremely variable within countries, owing to differences in land rental values, land quality and the amount of land used for dairy production.

- *High Land Costs (more than US$5/100 kg of milk):* High land costs were observed in Morocco (MA-12), New Zealand (NZ-282), Peru (PE-6) and Viet Nam (VN-2).
- *Intermediate land costs (US$2-5/100 kg of milk):* The majority of farms had land costs falling into this range and representative cases were found in Bangladesh, Cameroon, Germany, India (Karnataka, Orissa, Punjab), Morocco, New Zealand, Peru, Uganda, and the USA.

Low Land Costs (less than US$2/100 kg of milk): Low land costs were estimated for some farm types in Bangladesh, Cameroon, India (Karnataka, Orissa, Punjab), Pakistan, Thailand, the USA and Viet Nam.

Land Rental Prices

In the countries reviewed, land rents range from US$9 to US$745 per hectare:

- *More than US$300/ha:* All farms in Morocco, New Zealand and Viet Nam; and large farms in Germany and India (Punjab).
- *US$100 to 300/ha*: All farms in Bangladesh, India (Orissa and Karnataka), Pakistan, Peru and the USA; and the small farm type in Germany.
- *Less than US$100/ha*: All farm types in Cameroon and Thailand.

The level of land rental prices in Viet Nam can be explained by high population density in the areas where they were located (in the vicinity of Hanoi) and the large demand for urban/peri-urban land.

Land Productivity

Land productivity ranges from 1 000 to 154 000 kg of milk/ha. These differences are mainly a result of the various types of milk production system adopted, which differ from country to country and even within countries. For instance, farms in India (IN-6OR and IN-4KA), Thailand (TH-106) and Viet Nam (VN-4) have a very high milk output per area of farmland, which is mainly determined by large purchases of feed and fodder, and intensive use of concentrates.

Capital Costs

It is often difficult to estimate capital costs incurred in dairy farming because any cost comparison is challenged by differences in depreciation schemes, land valuation, levels of inflation, capital structure and interest rates. A method that yields sufficiently accurate results for most cases is to calculate interest on liabilities at 6 percent and for own capital at 3 percent. For some countries, among them China, this method slightly overestimates capital costs; for others, including India and Pakistan, the method produces slight underestimates.

Capital costs contribute approximately 3 to 29 percent of the total costs of the dairy farms studied, and thereby explain more or less as much as land costs on total cost differences. Capital costs range from US$0.8 to US$5.7/100 kg of milk, but for most farms they are in the order of US$1 to 2/100 kg of milk.

- *High capital costs* (more than US$3/100 kg) were observed for all Moroccan, New Zealand and Ugandan dairy farm types and for the small farm types in Cameroon, Germany and India.
- *Low capital costs* (less than US$1/100 kg of milk) were observed for the larger farm types in China and in Pakistan.

The considerable capital costs in New Zealand can be explained by high land prices and cost of shares in the cooperatives.

Capital Input/Cow

Capital costs per 100 kg of milk are a cost component of dairy production but do not provide information on the capital intensity of a dairy farm because, to a large extent, they are determined by the level of farm liabilities. The indicator 'capital input/cow', which only takes account of the capital embedded in the dairy herd, buildings and machinery and cooperative shares, gives a better sense of the 'capital-intensity' of a dairy farm. The different levels observed on the farms under study were as follows:

— *High (more than US$2 500/cow)*: All farms in Germany, Morocco, New Zealand and the USA.

Intermediate (US$1 000 to 2 500/cow): All farm types in Thailand and Viet Nam; the large farm type in Peru and the small one in Cameroon.

Low (less than US$1 000 US$/cow): All farm types in Bangladesh, China, India, Pakistan and Uganda; the large farm type in Cameroon and the small one in Peru.

References

Arnold, Wayne (2007). "In a growing world, milk is the new oil". *The New York Times*.

Evans, Gavin (2008). "N.Z. Forecasts Fall in Dairy Prices Through 2009". *Bloomberg.com.* Retrieved 2008-09-26.

Fraser, A.F. and D.M/ Broom. (1990). F*arm Animal Welfare and Behaviour* (3rd ed.) London: Bailliere Tindall. pp.355-356.

Greenough, P.R. (2007). *Bovine Laminitis and Lameness: A Hands-On Approach.* Edinburgh: Saunders. p.3.

Rushen, J., et.al.. (2008). The welfare of cattle. *Animal Welfare* Vol. 5. Berlin: Springer Verlag. pp. 21-35.

7

Poultry Farming

Keeping poultry makes a substantial contribution to household food security throughout the developing world. It helps diversify incomes and provides quality food, energy, fertilizer and a renewable asset in over 80 percent of rural households.Small-scale producers are however constrained by poor access to markets, goods and services; they have weak institutions and lack skills, knowledge and appropriate technologies. The result is that both production and productivity remain well below potential and losses and wastage can be high. However, adapted breeds, local feed resources and appropriate vaccines are available, along with proven technologies that can substantially improve productivity and income generation.

Importance of Poultry Farming

Family poultry is defined as small-scale poultry keeping by households using family labour and, wherever possible, locally available feed resources. The poultry may range freely in the household compound and find much of their own food, getting supplementary amounts from the householder. Participants at a 1989 workshop in Ile-Ife, Nigeria, defined rural poultry as a flock of less than 100 birds, of unimproved or improved breed, raised in either extensive or intensive farming systems. Labour is not salaried, but drawn from the family household. Family poultry was additionally clarified as "small flocks managed by individual farm families in order to obtain food security, income and gainful employment for women and children". Family poultry is quite distinct from medium to large-scale commercial poultry farming.

Family poultry is rarely the sole means of livelihood for the family but is one of a number of integrated and complementary farming activities contributing to the overall well-being of the household. Poultry provide a major income-generating activity from the sale of birds and eggs. Occasional consumption provides a valuable source of protein in the diet. Poultry also play an important socio-cultural role in many societies. Poultry keeping uses family labour, and women (who often own as well as look after the family flock) are major beneficiaries. Women often have an important role in the development of family poultry production as extension workers and in vaccination programmes.

For smallholder farmers in developing countries (especially in low income, food-deficient countries [LIFDC]), family poultry represents one of the few opportunities for saving, investment and security against risk. In some of these countries, family poultry accounts for approximately 90 percent of the total poultry production. In Bangladesh for example, family poultry represents more than 80 percent of the total poultry production, and 90 percent of the 18 million rural households keep poultry. Landless families in Bangladesh form 20 percent of the population and they keep between five and seven chickens per household. In LIFDC countries, family poultry-produced meat and eggs are estimated to contribute 20 to 30 percent of the total animal protein supply , taking second place to milk products (38 percent), which are mostly imported.

Similarly, in Nigeria, family poultry represents approximately 94 percent of total poultry keeping, and accounts for nearly four percent of the total estimated value of the livestock resources in the country. Family poultry represents 83 percent of the estimated 82 million adult chickens in Nigeria. In Ethiopia, rural poultry accounts for 99 percent of the national total production of poultry meat and eggs.

Poultry are the smallest livestock investment a village household can make. Yet the poverty-stricken farmer needs credit assistance even to manage this first investment step on the ladder out of poverty. Poultry keeping is traditionally the role of women in many developing countries. Female-headed households represent 20 to 30 percent of all rural households in Bangladesh, and women are more disadvantaged in terms of options for income generation. In sub-Saharan Africa, 85 percent of all households keep poultry, with women owning 70 percent of the poultry.

Income generation is the primary goal of family poultry keeping. Eggs can provide a regular, albeit small, income while the sale of live birds provides a more flexible source of cash as required. For example, in the Dominican Republic, family poultry contributes 13 percent of the income from animal production. The importance of poultry to rural households is illustrated by the example below from the United Republic of Tanzania. Assuming an indigenous hen lays 30 eggs per year, of which 50 percent are consumed and the remainder have a hatchability of 80 percent, then each hen will produce 12 chicks per year. Assuming six survive to maturity (with 50 percent mortality), and assuming that three pullets and three are cockerels, the output from one hen projected over five years would total 120 kg of meat and 195 (6.8 kg) eggs.

A study on income generation in transmigrant farming systems in East Kalimantan, Indonesia, showed that family poultry accounted for about 53 percent of the total income, and was used for food, school fees and unexpected expenses such as medicines.

Flock composition is heavily biased towards chickens in Africa and South Asia, with more ducks in East Asia and South America. Flock size ranges from 5 - 100 in Africa, 10 - 30 in South America and 5 - 20 in Asia. Flock size is related to the poultry farming objectives of:

- home consumption only;

- home consumption and cultural reasons;
- income and home consumption; and
- income only.

In Bangladesh, the average production rate per local hen of 50 eggs/year was regarded by some as low productivity. However, if it is considered that 50 eggs per hen per year represents four hatches from four clutches of eggs laid, incubated and hatched by the mother hen, and the outcome is 30 saleable chicken reared per year (assuming no eggs sold or eaten, 80 percent hatchability and 25 percent rearing mortality), then it is a remarkably high productivity.

Production Systems

Family poultry are kept under a wide range of conditions, which can be classified into one of four broad production systems:

- free-range extensive;
- backyard extensive;
- semi-intensive; and
- intensive.

Free-Range Extensive Systems

In Africa, Asia and Latin America, 80 percent of farmers keep poultry in the first two extensive systems. Under free-range conditions, the birds are not confined and can scavenge for food over a wide area. Rudimentary shelters may be provided, and these may or may not be used. The birds may roost outside, usually in trees, and nest in the bush. The flock contains birds of different species and varying ages.

Backyard Extensive Systems

Poultry are housed at night but allowed free-range during the day. They are usually fed a handful of grain in the morning and evening to supplement scavenging.

Semi-Intensive Systems

These are a combination of the extensive and intensive systems where birds are confined to a certain area with access to shelter. They are commonly found in urban and peri-urban as well as rural situations. In the *"run"* system, the birds are confined in an enclosed area outside during the day and housed at night. Feed and water are available in the house to avoid wastage by rain, wind and wild animals.

In the European system of free-range poultry keeping, there are two other types of housing. The first of these is the *"ark"* system, where the poultry are confined overnight (for security against predators) in a building mounted on two rails or skids (usually wooden), which enable it to be moved from place to place with draught power. A typical size is 2 × 2.5 m to hold about 40 birds.

The second type of housing is the *"fold"* unit, with a space allowance (stock density) for adult birds of typically 3 to 4 birds per square metre (birds/m^2), both inside and (at least this) outside. The fold unit is usually small enough to be moved by one person. Neither of these two systems is commonly found in developing countries.

Intensive Systems

These systems are used by medium to large-scale commercial enterprises, and are also used at the household level. Birds are fully confined either in houses or cages. Capital outlay is higher and the birds are totally dependent on their owners for all their requirements; production however is higher. There are three types of intensive systems:

- *Deep litter system:* birds are fully confined (with floor space allowance of 3 to 4 birds/m^2 within a house, but can move around freely. The floor is covered with a *deep litter* (a 5 to 10 cm deep layer) of grain husks (maize or rice), straw, wood shavings or a similarly absorbent (but non-toxic) material. The fully enclosed system protects the birds from thieves and predators and is suitable for specially selected commercial breeds of egg or meat-producing poultry (layers, breeder flocks and broilers).
- *Slatted floor system*: wire or wooden slatted floors are used instead of deep litter, which allow stocking rates to be increased to five birds/m^{2}of floor space. Birds have reduced contact with faeces and are allowed some freedom of movement.
- *Battery cage system*: this is usually used for laying birds, which are kept throughout their productive life in cages. There is a high initial capital investment, and the system is mostly confined to large-scale commercial egg layer operations.

Intensive systems of rearing indigenous chickens commercially is uncommon, a notable rare exception being in Malaysia, where the industry developed in response to the heavy demand for indigenous chickens in urban areas. However, this accounts for only two in every 100 000 (0.002 percent) of that country's indigenous chicken.

Poultry Species and Breeds

All species of poultry are used by rural smallholders throughout the world. The most important species in the tropics are: chickens, guinea fowl, ducks (including Muscovy ducks), pigeons, turkeys and geese. Local strains are used, but most species are not indigenous. The guinea fowl (*Numididae)* originated in West Africa; the Muscovy duck (*Cairina moschata*) in South America; pigeons *(Columba livea*) in Europe; turkeys*(Meleagrididae)* in Latin America; pheasants (*Phasianidae*) in Asia; the common duck *(Anas)* in Europe; and geese (*Anser*) in Asia.

Flock composition is determined by the objectives of the poultry enterprise. In Nigeria for example, the preference is for the smooth-feathered, multicoloured native chickens or Muscovy ducks. Multicoloured feathers serve as camouflage for scavenging birds against predators, including birds of prey, which can more easily see solid colours (especially white). Foundation stock is usually obtained from the market as grower pullets and young cockerels. A hen to cock ratio of about 5:1 is common. Both sexes are retained for 150 to 300 days, for the purposes of culling, selling, home consumption and gifts, most of which require adult birds.

In the last 50 years, there has been a great advance in the development of hybrid breeds for intensive commercial poultry production. This trend is most noticeable in chickens, turkeys and ducks. The new hybrids (those of chickens in particular) are widely distributed and are present in every country in the tropics, even in the most remote villages. The hybrids have been carefully selected and specialised solely for the production of either meat or eggs. These end-product-specialised hybrid strains are unsuitable for breeding purposes, especially for mixing with local village scavenger stock, as they have very low mothering ability and broodiness.

For the smallholder, keeping hybrids means considerable changes are required in management. These changes are expensive for the following reasons:

- All replacement day-old chicks must be purchased.

- Hatchery chicks require artificial brooding and special starting feed.
- Hybrids require higher quality balanced feed for optimum meat and egg production.
- Hybrids require more careful veterinary hygiene and disease management.
- Egg-laying hybrid hens require supplementary artificial light (a steadily increasing day-length up to 17 hours of total light per day) for optimum (profitable) egg production.

The meat and eggs from intensively raised hybrid stock are considered by many traditional consumers to have less flavour, and the meat to have too soft a texture. Consumers will thus often pay a higher price for village-produced poultry meat and eggs. Thus for rural family poultry keepers, it is more appropriate to maintain and improve local birds to meet this demand.

Chickens

Chickens originated in Southeast Asia and were introduced to the rest of the world by sailors and traders. Nowadays, indigenous village chickens are the result of centuries of cross-breeding with exotic breeds and random breeding within the flock. As a result, it is not possible to standardize the characteristics and productive performance of indigenous chickens.

There is no comprehensive list of the breeds and varieties of chickens used by rural smallholders, but there is considerable information on some indigenous populations from various regions. Most of this is based on feather colour and other easily measured body features (genetic traits), but more detailed data are becoming available.

Characteristics such as adult body weight and egg weight vary considerably among indigenous chicken populations, although reproductive traits, such as the number of laying seasons per year, the number of eggs per clutch and hatchability are more consistent. *Desi* hens in Bangladesh range from 190 to 200 days of age at first egg (an easy measure of age-at-sexual-maturity), and they lay 10 to 15 eggs per season in 3 to 4 clutches (3 to 4 times) per year, with a hatchability of 84 to 87 percent (percent of eggs set).

Indigenous village birds in Ethiopia attain sexual maturity at an average age of seven months (214 days). The hen lays about 36 eggs per year in three clutches of 12 to 13 eggs in about 16 days. If the hen incubates her eggs for three weeks and then rears the chicks for twelve weeks, then each reproductive cycle lasts for 17 weeks. Three cycles then make one year. These are very efficient, productive and essential traits for survival.

Guinea Ffowl

Guinea fowl are native to West Africa but are now found in many parts of the tropics, and are kept in large numbers under intensive systems in France, Italy, the former Soviet Union and Hungary. In India, guinea fowl are raised in parts of the Punjab, Uttar Pradesh, Assam and Madhya Pradesh, usually in flocks of a few hundred birds. Guinea fowl are seasonal breeders, laying eggs only during the rainy season,under free-range conditions. They are very timid, roosting in trees at night, and although great walkers, they fly very little.

Guinea fowl thrive in both cool and hot conditions, and their potential to increase meat and particularly egg production in developing countries deserves better recognition. The first egg is normally laid at about 18 weeks of age, and unlike many indigenous birds (which produce a single clutch a year), guinea hens lay continuously until adverse weather sets in. In West Africa, laying is largely confined to the rainy season. Guinea hens under free-range conditions can lay up to 60 eggs per season, while well-managed birds under intensive management can lay up to 200 eggs per year. The guinea hen "goes broody" (sits on eggs in the nest) after laying, but this can be overcome by removing most of the eggs. A clutch of 15 to 20 eggs is common, and the incubation period for guinea fowl is 27 days. Domesticated guinea fowl under extensive or semi-intensive management in Nigeria were reported to lay 60 to 100 eggs with a fertility rate of 40 to 60 percent.

Domesticated guinea fowl are of three principal varieties: Pearl, White and Lavender. The Pearl is by far the most common. It has purplish-grey feathers regularly dotted or "pearled" with white. The White guinea fowl has pure white feathers while the Lavender has light grey feathers dotted with white. The male and female guinea fowl differ so little in appearance (feather colour and body weight [1.4 to 1.6 kg]) that the inexperienced farmer may unknowingly keep all males or all females as "breeding" stock. Sex can be distinguished at eight weeks or more by a difference in their voice cry.

Domesticated guinea hens lay more eggs under intensive management. French Galor guinea hens can produce 170 eggs in a 36-week laying period. For example, from a setting of 155 eggs, a fertility rate of 88 percent and hatchability of 70 to 75 percent, it is possible to obtain 115 guinea keets (chicks) per hen. In deep litter or confined range conditions, a 24-week laying period can produce 50 to 75 guinea keets per hen.

Ducks

Ducks have several advantages over other poultry species, in particular their disease tolerance. They are hardy, excellent foragers and easy to herd, particularly in wetlands where they tend to flock together. In Asia, most duck production is closely associated with wetland rice farming, particularly in the humid and subtropics. An added advantage is that ducks normally lay most of their eggs within the three hours after sunrise (compared with five hours for chickens). This makes it possible for ducks to freely range in the rice fields by day, while being confined by night. A disadvantage of ducks (relative to other poultry), when kept in confinement and fed balanced rations, is their high feed wastage, due to the shovel-shape of their bill. This makes their use of feed less efficient and thus their meat and eggs more expensive than those of chickens. Duck feathers and feather down can also make an important contribution to income. Different breeds of ducks are usually grouped into three classes: meat or general purpose; egg production; and ornamental.

Ornamental ducks are rarely found in the family poultry sector. Meat breeds include the Pekin, Muscovy, Rouen and Aylesbury. Egg breeds include the brown Tsaiya of Taiwan Province of China, the Patero Grade of the Philippines, the Indian Runner of Malaysia and the Khaki Campbell of England. All these laying breed ducks originate from the green-headed Mallard. The average egg production of the egg breeds is approximately 70 percent (hen.day basis). The Indian Runner, Khaki Campbell, Pekin and Muscovy are the most important breeds in rural poultry.

Geese

Geese are less important in family poultry production, except in China, where mainly local breeds are kept, except for a few European breeds such as the Toulouse and White Roman, imported for cross-breeding purposes. The great variety in breed size of geese permits their use under various management conditions. At the less intensive levels of production preferred by most family producers, smaller-sized birds (weighing approximately 4 kg, such as the Lingxhian or Zie breeds in China) are easier to manage. Geese are high in the broodiness trait, and have a consequent low egg production of 30 to 40 hatching eggs (in three to five laying cycles) per year. At the other extreme are breeds of high fertility (and egg number), which are smaller and are selected specifically for use in breeding flocks for their lack of broodiness. Breeds such as the Zie may lay 70 to 100 eggs annually. The importance of the wide gene pool variety in China is significant for the Asian region in particular and for the world in general.

Pigeons

Pigeons are scavengers (not fed any supplementary feed) in most countries, living on the roofs of houses and treated as "pets" that do not need to be fed. They appear to prefer homestead compounds to fields. In some countries, they are eaten only for ritual purposes. They normally lay two eggs in a clutch, and the young birds (squabs) hatch after 16 to 17 days. The growing squabs are fed by their mothers on *crop milk*, produced in the mother's crop (first stomach). This enables young squabs to grow very rapidly. They reach maturity in three to five months at a body weight of 200 to 300 g for males, and 150 g for females. Adult pigeons are monogamous for life.

Local pigeons are specific to different regions in the tropics. Africa has five breeds, within which Chad has three local breeds. Asia and the Pacific have five breeds, with local breeds found specific even to the Cook Islands. Latin America and the Caribbean islands have only one breed. Europe has six breeds, two of which come from Belgium.

Turkeys

These birds are native to Latin America. The breeds kept by rural producers in the tropics usually have black feathers, as distinct from the white-feathered breeds that are raised intensively. Where there are no geese and ostriches, they are the largest birds in the farming system. Body weight ranges from 7 to 8 kg in males and from 4 to 5 kg in hens. They have good meat conformation, produce about 90 eggs per year and have medium to good hatchability. They are more susceptible to disease than either chicken or ducks.

Feed Resources

A regular supply of low-cost feed, over and above maintenance requirements, is essential for improved productivity in the three farming systems used in family poultry production:

- *free-range* - poultry roost in trees at night;
- *backyard* - poultry are confined at night; and
- *semi-intensive* - poultry are enclosed during the day in a very limited scavenger resource base.

When feed resources are inadequate, a few birds in production are better than more birds just maintained, but without enough food for production.

Extensive Systems

Farmers attempt to balance stock numbers according to the scavenging feed resources available in the environment in each season. Under the free-range and backyard systems, feed supplies during the dry season are usually inadequate for any production above flock-maintenance level. When vegetation is dry and fibrous, the scavenging resources should be supplemented with sources of minerals, vitamins, protein and energy. Under most traditional village systems, a grain supplement of about 35 g per hen per day is given.

There have been various approaches to utilising a wider base of feed resources for the flock. One is the use of poultry species apart from chicken. Waterfowl, especially ducks, may be distributed throughout the wetland rural areas, where they can feed on such resources as snails and aquatic plants in ponds and lagoons. Another approach is the integration of poultry with the production of rice, vegetables, fish and other livestock. An example is the combination of chicken with cattle, as practised by the Fulani of Nigeria, where the chickens feed on the ticks on the cattle as well as on the maggots growing in the cattle dung. Chickens raised in the cattle kraal (compound) weighed an average of 500 g more than those in the same neighbourhood but outside the kraal.

Semi-Intensive System

Under the semi-intensive system, all the nutrients required by the birds must be provided in the feed, usually in the form of a balanced feed purchased from a feed mill. As these are often expensive and difficult to obtain, smallholders use either unconventional feedstuffs or "dilute" the commercial feed by supplementing it with grain by-products. A well-balanced feed however is difficult to achieve, as grains and plant protein sources are becoming increasingly unavailable for livestock, and premixed trace minerals and vitamins are usually too expensive for smallholders. Phosphorus and calcium can be obtained from ashed (burnt and crushed) bones; and calcium from snail shells, fresh or seawater shell shells, or limestone deposits. Salt to supply sodium can come from evaporated seawater or land-based rock salt deposits. These mineral sources are rarely used. Feed provided for birds kept under this system is therefore of a much poorer quality than under either the extensive or fully intensive system.

General Management

Housing and Runs

Under undomesticated conditions, poultry lay eggs in simple nests, perch in trees and spend much of the day scavenging for feed. Chickens spend a large proportion of their time scratching to expose hidden food. Under the backyard and semi-intensive production systems, poultry are usually enclosed at night to discourage thieves and predators, and under intensive production, are totally confined day and night. Some village households keep their few chickens inside the house or even under their bed at night, to discourage theft.Given a choice of a place to lay their eggs, hens will choose a soft "litter" base, and they prefer an adequately sized, darkened nest with some privacy. Prior to laying, hens usually investigate a number of possible sites before entering a nest box. They then show nesting behaviour, which includes a special protective nest-seeking voice, after which they sit and finally lay. When they have laid an egg, they announce this with another type of "pride of achievement" call. These calls can also be heard in a battery cage house. If perches are provided, hens will perch most of the time rather than stand on the wire floors, and after dark most birds roost on the perches. Perching is a probable survival characteristic to avoid night predators.

Many factors influence the type and choice of housing to protect poultry from the effects of weather and predators. These include the local climate, the available space, the size of the flock and the management system. In extensive systems, birds must be protected from disease and predators but also be able to forage. Traditional large animal fencing using live plants is not enough protection against predators such as snakes, kites, rats and other vermin.A simple and effective system to deter predator birds is to tie parallel lines of string across the main scavenging area, the intervals between which measure less than the predator's wingspan; or, alternatively, a ing net supported on poles can be spread across the side of the run where predator birds could swoop on the scavenging chicks.

The detailed description of poultry housing is given in Chapter 2 (pp.65-76)

Baby Chick Management

Baby chicks should be kept warm and dry. The nest, which they share at night with the mother hen, must be kept clean. In colder climates (below 20 °C at night), the nest site should be kept warm by lining it with straw and placing it near a stove or fireplace. The chicks should remain with the mother hen for nine to ten weeks, learning from her example how to scavenge and evade predators and other dangers. Clean drinking water and fresh feed in a clean container should be provided to supplement scavenging.

There is a close relationship between chick weight and growth and mortality rates. In an experiment where young chickens had access to supplementary feed in a creep feeder, it was found that supplementary protein feed had a significant effect on the survival rate and growth rate. Chicks separated from the mother hens during the day from the age of three to ten weeks, and fed with chicken starter mash *ad libitum*, had a mortality rate of 20 percent and a body weight of 319 g at ten weeks, compared with a mortality rate of 30 percent and a body weight of 242 g for the control group which remained with the mother hens. A suitable strategy for rearing chicks therefore would be as follows:

- The chicks should be confined for the first weeks of life and provided with a balanced feed.
- A vaccination programme should be followed.
- Sufficient supplementary feed should be provided during the remaining rearing period to allow the chickens to develop in accordance with their genetic potential.
- Feed supplements and protection should be provided to naturally brooded chickens during the first four to eight weeks of life.

The composition of the supplementary feed will depend on the available scavengable feed, but a form of cafeteria free-choice feeding of a protein concentrate, energy concentrate and calcium mineral in each of three containers may be the best solution.

The mortality rate of naturally brooded chicks, whose only source of feed is from scavenging under free-range conditions, is very high and often exceeds 50 percent up to eight weeks of age. Wickramenratne *et al* found that predators accounted for up to 88 percent of mortality and that coloured birds had a higher survival rate than white birds. The high mortality rate and the large number of eggs required for hatching are the main causes of low offtake from scavenging poultry flocks. Smith reported an offtake (sales and consumption) of only 0.3 chickens per hen/year from a survey done on flocks in Nigeria. This low offtake has also been observed in Bangladesh and India.

An efficient way of decreasing mortality rate (a costly loss) is to confine and vaccinate the chicks during the rearing period. This however is more expensive, the cost of feed in particular increasing production costs. A method used over the past ten years in many poultry development projects in Bangladesh confines the chicks during the first eight weeks of life. They are fed approximately 2 kg each of balanced feed and thereafter kept under semi-scavenging conditions. At eight weeks of age, they are less susceptible to attacks by predators and more resistant to diseases, due to their larger body weight and more effective vaccination immunization (due to their better nutrient intake).

Manure Management

Whatever the type of confinement, proper attention must be paid to manure management. Adult birds produce 500 g of fresh manure (70 percent moisture content) per year per kg of body weight. To preserve its fertilizer value, manure should be dried to about 10 to 12 percent moisture content before storage. This will retain the maximum nitrogen content for fertilizer value. Nitrogen in the form of urea is the most volatile component of manure, and is lost as ammonia if moisture content is too high in the stored material. If the moisture content is too high, then the stored manure releases ammonia, carbon dioxide, hydrogen sulphide and methane, which can have serious physiological effects on humans. Some of these components are also greenhouse gases, which contribute to the global increase in ambient temperature. Poultry manure is very useful as an organic fertilizer, as animal and fish feed and as a raw material for methane gas generation in biogas plants for cooking fuel.

Management of Free-range Poultry

The unrestricted free-ranging of poultry is often a problem. They trespass onto neighbouring fields and gardens, and are constantly at risk from predators. Confinement is often not practical because of the cost of feed and fencing, while surveillance is only feasible where the very old or very young of the household have time to help. Fencing of vegetable plots is in many cases the best option. Placing more cocks in the village might reduce the movements of the chickens, as the cocks and hens of each flock would keep more to their own territory. Cocks move within an eight-to-ten-house territory, and hens within two or three houses.

Under the free-range system, the difference between the amounts of food gathered through scavenging and the total food requirement for maximum production should be balanced with nutrients supplied from supplementary feed. To make up a properly balanced supplement, it is necessary to know the scavenger feed resource base (SFRB) and the composition of the crop contents. If this is not known, it is recommended that the fowls have access (using a free-choice cafeteria system) to three containers (or three compartments of a bamboo stem feeder of ingredients comprising a protein concentrate, a carbohydrate source (for energy) and a mineral source (mainly for calcium carbonate for egg shell formation for the hen). Poultry should have free access to this cafeteria system for two to three hours in the evening to supplement the day's scavenging.

From a feed resource point of view, this recommendation is only economically viable (sustainable) if the consumption of supplementary feed per egg produced is equal to 150 to 180 g or less. Consumption of over 150 g is only justified if the supplements are cheaper than the commercial feed used in intensive poultry production. Supplements are usually recommended in the range of 50 to 80 g/bird/day, so it is usually quite viable. Seasonal variations in the SFRB have a substantial effect on production. During the dry season, scavenged feed from gardens, crops and wasteland (such as grass shoots, seeds, worms, insects and snails) stops, while the quantity and quality of household kitchen waste decrease. The feed supplement should be adjusted seasonally to maintain an optimum level of production or, alternatively, the chicken population could be adjusted to the amount of the SFRB and the feed supplement.

Hens in confinement fed a balanced diet will convert food weight to egg weight at an efficiency of about 2.8 kg of feed per kilogram of egg weight. Changes in husbandry alone may increase the productivity of scavenging village chickens, without the need for additional inputs. In planted orchards, a stocking rate of 120 to 180 birds/ha will clean up windfalls while also fertilizing the trees. In this example, the amount of fertilizer produced per hectare for 150 hens (weighing two kilograms each) is based on the assumption of 500 g of fresh (70 percent moisture) weight of manure produced per kilogram of live weight per year. This results in 330 g of dried manure (dried to a ten percent moisture content) per hen/year, and thus the 150 hens will produce 49.5 kg of dry manure per year. This has an equivalent fertiliser value of 13 percent ammonium nitrate, 8.6 percent super-phosphate and 2.9 percent potash (potassium) salts. Thus the 150 hens will produce the equivalent per hectare/year of 6.4 kg of ammonium nitrate, 4.3 kg of super-phosphate and 1.4 kg of potash salts.

Planning Flock Production and Size

Production involves birds for meat and eggs. For both meat and egg production, the number of chickens in the flock is the most important factor. Flock size changes constantly as eggs hatch and hens are sold or eaten. Usually the main cause of flock depletion is mortality, particularly in chicks. Disease is the greatest cause of mortality, especially in the rainy season and in the weather changeable humid periods on either side. During summer and the rainy season, predators in the cropped fields also contribute to reduced flock sizes. Local birds lay an average of three to four clutches of 12 to 15 eggs in a year, with more eggs laid at crop harvest time because more feed is available. Given most traditional farming systems, keeping the flock number constant requires eight to ten eggs for reproduction, leaving an average of 35 to 40 eggs per layer for sale or consumption. Because the number of eggs needed for replacement may decrease with better management, the extra eggs can be sold or eaten.

Most egg laying takes place between sunrise to mid-morning. During the months of laying, nest location should not be moved, as this may upset the laying routine.

In village flocks, income derives from the sale of eggs and live birds. For example, a flock of 15 local hens laying 30 eggs/hen/year (with one local cock) will produce 450 eggs in a year. Of these 450 eggs, 120 may be incubated by broody hens (in ten clutches of 12 eggs each), of which 100 chicks may hatch, and 30 eggs may be cracked and consumed in the household, leaving a balance of 300 eggs for sale. Of the 100 day-old chicks, 30 may reach maturity (with rearing losses of 70 percent), to yield 15 cockerels and 15 pullets. The 15 pullets will replace the older hens, of which ten remain after the sale of cull hens, and one new cockerel will replace the old cock. The annual income from the flock can therefore be calculated as follows:

300 eggs + 10 old hens + 1 old cock + 14 cockerels = income

For improved productivity, culling is important and productive birds should be carefully selected. For simplicity, the above example assumes no adult mortality.

Incubation and Hatching

Natural Incubation

The broody hen chosen for natural incubation should be large (to cover and thus keep more eggs warm), healthy and preferably vaccinated, with a good brooding and mothering record. Signs of broodiness are that the hen stops laying, remains sitting on her eggs, ruffles her feathers, spreads her wings and makes a distinctive clucking sound. Brooding may be induced with dummy eggs or even stones.

Eggs usually become fertile about four days after the rooster has been introduced to the hens. A maximum of 14 to 16 eggs may be brooded in one nest, but hatchability often declines with more than ten eggs, depending on the size of the hen. Feed and water provided in close proximity to the hen will keep her in better condition and reduce embryo damage due to the cooling of the eggs if she has to leave the nest to scavenge for food.

The hen keeps the eggs at the correct humidity by splashing water on them from her beak. This is a further reason for providing her with easy access to water. In very dry regions, slightly damp soil can be placed under the nesting material to assist the hen in maintaining the correct humidity. Fertile eggs from other birds are best added under the brooding hen between one and four days after the start of brooding. In Bangladesh, it has been reported that local broody hens will even sit on and hatch a second clutch of eggs, often losing considerable weight in the process .

The incubation period for chicken eggs is 20 to 21 days, and increases up to 30 days for other poultry. After sitting for some days, a broody hen can be given some newly hatched chicks and, if they are accepted, the original eggs can be removed and replaced with more chicks. Thus hens with a better record of mothering can be better utilised for their abilities.

Eggs initially need a very controlled heat input to maintain the optimum temperature of 38°C, because the embryo is microscopic in size. As the embryo grows in size (especially after 18 days), it produces more heat than it requires and may even need cooling. Moisture levels of 60 to 80 percent Relative Humidity are important to stop excess moisture loss from the egg contents through the porous egg shell and membranes. Factors to consider for successful natural incubation include the following:

- Feed and water should be close to the hen.
- The broody hen should be examined to ensure that she has no external parasites.
- Any eggs stored for incubation should be kept at a temperature between 12 and 14 °C, at a high humidity of between 75 to 85 percent, and stored for no longer than seven days.
- Extra fertile eggs introduced under the hen from elsewhere should be introduced at dusk.
- The eggs should be tested for fertility after one week by holding them up to a bright light (a candling box works best. If there is a dark shape inside the egg (the developing embryo), then it is fertile. A completely clear (translucent) egg is infertile.

A hatchability of 80 percent (of eggs set) from natural incubation is normal, but a range of 75 to 80 percent is considered satisfactory. Setting of hatchings is best timed so that the chicks to be hatched are two months of age at the onset of major weather changes, such as either the rainy (or dry) season or winter/summer. A plentiful natural food supply over the growing period of the chicks will ensure a better chance for their survival. Successful poultry species instinctively lay and incubate their eggs at a time of the year when newly hatched chicks will have a better supply of high protein and energy food provided by the environment. For example, guinea fowl will only lay eggs in the rainy season. However, seasonal changes in weather patterns are also times of greater disease risk.

Artificial Incubation

There are many commercial artificial incubators of varying capacities. Most depend on electricity, but some use gas or kerosene for heating. All use a thermostatic switching device to keep the temperature constant within one Celsius degree. The correct humidity is usually maintained by having a pre-determined surface area of water appropriate for each incubator chamber.

Turning the egg several times each day is important to prevent the embryo from sticking to the shell membranes. With hand-turning systems, an odd number of times turned per day (five to seven times) will ensure that during successive overnight periods the egg is always oriented the opposite way from that of the previous night.

The broody hen carries out all of these incubation tasks instinctively, and artificial incubation attempts to duplicate these tasks. Traditional artificial incubation techniques have evolved over thousands of years in many parts of the world. One such technique, developed for hatching duck eggs in China, is the parched (heated) rice technique. It is based on the use of heated paddy rice and embryo-generated heat. It is still used in China and Bali, Indonesia, with hatchability results of up to 80 percent). The objectives of artificial incubation are met equally well using either parched rice or rice husks, and a hatchability of 65 to 75 percent is common. By candling the eggs between days 5 and 7, infertile eggs can be detected as "clears" (as the light is not obscured by the growing embryo). These eggs are still suitable for sale for human consumption, which improves the economic viability of this system.

As duck eggshells are less brittle than chicken eggs, the system was never adopted for chicken eggs in China. The original Chinese system used 80 duck eggs per bundle. However, with extra care, and fewer eggs per bundle (25 to 30 compared with 40 duck eggs), chicken egg incubation was found to be equally successful in Bangladesh when adapted there in the 1980s. The number of duck eggs per bundle was reduced to 40, which gave better hatching results as well as fewer breakages than 80 per bundle.

The artificial parched rice or rice husk incubation process is started by heating the eggs, either in the sun or in an insulated warming room equipped with a heat source. On sunny days, approximately 25 to 30 chicken eggs or 40 duck eggs, (presumed fertile, and carefully dated and labelled) are placed in the sun on pieces of padded cloth for about 30 minutes and turned occasionally to raise the temperature of the eggs to the required 37 to 38 °C. This temperature can be judged by the appearance of water droplets on the shell or by touching the egg to the eyelid. On sunless days, eggs must be placed on a cloth in a shallow bamboo basket and put on racks in a heated warming room to slowly achieve the same temperature. This usually requires approximately one to three hours. Any slow-burning fuel is suitable, and kerosene and charcoal are commonly used. In Egypt, dried beanstalks were used for thousands of years for their characteristic slow-burning property, and are still used today in the Fayoumi district (from where the well-known chicken breed of the same name originates). A well-vented stove will prevent any toxic fumes affecting the embryo. In Bangladesh, slightly heated (to 38°C) sand or wood-ash, covering the eggs for approximately one minute, is also effective in warming the eggs. Human clinical thermometers, now readily available, can be used to assist training in using the eyelid's sensitivity to temperature. If the humidity drops below 70 percent, 60 percent, 50 percent or even 40 percent (particularly in typical progressively drier months), a wet (and slightly warm) cloth should be placed over the eggs with a frequency corresponding to the above humidities of one, two, four or six times daily. This will raise the humidity of the egg so that the embryo will not dry out.

The unhusked (paddy) rice is heated (parched) and continuously stirred until it reaches a temperature of 60°C, to provide heat for the eggs for the first two weeks of incubation. About 2.5 to 3.0 kg of the heated rice is enclosed in a cloth pillow and placed into the egg basket. The pillows have the same diameter (40 cm in the Bangladesh example) as the basket, and should be about 8 cm thick. Where rice-husk is used, pillows can be made with black-coloured material, which easily absorb the sun's heat. When the temperature of the pillow has dropped to about 40°C, a loosely bunched bundle of 40 duck (or 25 to 30 chicken) pre-warmed eggs are placed on top of the warm rice. The bundle is made from a square piece of cloth about 45 cm on each side. A soft duster cloth with pinholes is suitable. Alternating pillows of warm rice and egg bundles are added until the basket is full, finishing with a pillow of rice. The basket is then covered with padding to conserve as much heat as possible. This procedure is repeated until all the eggs are placed in baskets, leaving one basket empty to allow the addition of freshly warmed rice. The incubating baskets are cylindrical in shape (50 cm in diameter and 80 cm in depth in China, and 40 cm in diameter and 70 cm in depth in Bangladesh and are made of woven bamboo strips. In China, a few layers of bamboo paper are pasted on the inner surface of the sides and bottom to seal the gaps against temperature loss through convection.

In warmer parts of southern China, rice husk is substituted for paddy rice, and the pillows containing the husks are made from black-coloured material, which easily heat up in the sun. Rice husks also provide very good insulation against loss of heat from the older eggs, which is instead transferred (by conduction) to younger eggs placed in contact with them in separate bundles. This rice-husk system was adopted on a large scale in Bangladesh after being introduced by poultry development projects in the 1980s. The system has evolved, and the cylindrical egg baskets are now set into larger bamboo frame setting boxes, with more insulating rice husk material placed between the cylinders and the walls of the enclosing setting boxes. The cylinder wall should be about 10 cm from the setting box wall and 8 cm from the next cylinder. With this greater insulation, there is less heat loss, thus less need to provide supplementary heat from costly fuels.

For the first three days, reheated paddy rice (or rice husk) is added three times a day at regular intervals. During days four to six, this may be reduced to twice a day. The object is to ensure that the eggs are kept at the temperature most suitable for embryo development. The spare basket is used to transfer eggs from an adjoining basket when adding freshly warmed rice or rice-husks. Thus the top layer of eggs becomes the bottom layer and the bottom layer ends up on top of the spare cylinder. The newly emptied basket is then ready to receive eggs from the third basket, and so the cycle continues.

In China, heat is provided for days 13 to 14 in summer and in the colder months also on days 18 to 19. After that, the developing embryos are able to produce enough heat to maintain the incubation process without further need of an outside heat source. Eggs set for the first six days are called "new eggs", those between days 6 to 13 (which neither need nor produce extra heat) are "in-between eggs" and eggs after day 13 (which give out excess heat) are "old eggs". Once a basket contains "old" and "in-between eggs", it is possible to use embryo-generated heat alone to incubate "new eggs", which are usually introduced at intervals into the basket and placed between layers of "old eggs".

The eggs are candled on days 5 and 13, both to identify infertile eggs and dead embryos and to assess the degree of embryo development; which is used as one of the guides in adjusting basket temperature. Placing the egg on the upper eyelid allows the egg temperature to be assessed. The temperature of the basket may be adjusted in the following ways:

- by varying the proportion of eggs in a basket at different stages of incubation (for example, the temperature may be lowered by removing some of the bundles of "old eggs");
- by varying the arrangement of eggs in a basket (for example, as heat dissipation from eggs near the sides of the basket is faster than from those in the centre; a bundle of "old eggs" can be placed in the centre core and bundles of "new eggs" shaped to enclose them); and
- by changing the top covering of the basket using heavier padded material in cold weather and at the beginning of the incubation period, and a lighter covering if less heat retention is required.

Table 1: Arrangement of eggs in the incubation baskets prior to and after transfer to adjoining baskets

		Incubation time (days)		
Layer	*Before transfer*	*Summer*	*Winter*	*fter transfer*
1	A	14-6	17-20	J
2	B	1-3	1-4	I
3	C	10-13	13-16	H
4	D	4-6	5-8	G
5	E	7-9	9-12	F
6	F	7-9	9-12	E
7	G	4-6	5-8	D
8	H	10-13	13-16	C
9	I	1-3	1-4	B
10	J	14-16	17-20	A

Eggs in the advanced stages of incubation produce a lot of heat, so on days 13 to 14 in summer (days 18 to 19 in winter), the "old eggs" are transferred to hatching beds, where they are placed in a single layer for final development and hatching. The surface of the bed is covered with a thin layer of rice husks and then covered with a straw mat. The edges of the bed are lined with padding to protect the eggs. The covering for the developing eggs in the bed may be heavy or light cloth, depending on the degree of insulation required. The temperature in the hatching bed is maintained at 36 to 37 °C, slightly lower than that of the basket. The temperature can be adjusted by changing the thickness of the covering, varying the space between the eggs, and moving the eggs twice a day so that those on the perimeter change places with those at the centre. In very hot dry weather, the eggs are sprayed with a fine mist of water. They are kept in the bed until the chicks hatch out and dry.

The hatching bed should be two-storied (like a bed-bunk) and can be made of wood. An example hatching bed with a 500-egg capacity has a length of 90 cm and width of 68 cm. The height of the side wall on all sides should be 20 cm from the bed base level, with a 25 cm gap above the side wall for ventilation and ease of access.

Pests and Diseases of Poultry Flocks

In most family flocks, disease is an important problem. Although farmers are familiar with the signs and symptoms of disease, the underlying causes are less well known. Almost every farmer and most extension workers hold Newcastle Disease (ND) responsible for most deaths, and the disease has a local name in all languages.

Table 2: Signs of poultry disease observed by farmers in East Kalimantan villages, Indonesia

Signs	*Frequency%*
Chickens huddle together	16.1
Coughing, sneezing, rapid breathing	13.2
Discharge from mouth and nostrils	10.9
Dullness, no appetite, closed eyes	10.9
Paralysis of legs and wings	9.2
White droppings	8.6
Turned or twisted neck	8.0
Dark red colour of head and comb	6.9
Greenish or yellow droppings	4.6
Bloody reddish droppings	4.0
Swellings of head and comb	2.9
Pale comb	1.7
Worms in faeces	1.7
Eye worm	1.1

However, not all infectious diseases are due to the Newcastle Disease Virus (NDV). Digestive problems resulting in slow growth and diarrhoea may be the result of rancid feed or too much salt, and may also be symptoms of diseases such as coccidiosis, salmonellosis, or Gumboro Disease (Also called Infectious Bursal Disease [IBD]). ND often has the symptom of greenish faeces, which indicates a loss of appetite.

Poultry have highly developed and sophisticated respiratory systems. Their heart and breathing rate are faster than humans, and their body temperature is 5 °C higher. Their lungs are connected at the lower ends to a complex series of membrane-enclosed air sacs, which in turn are connected to air cavities in their major skeletal bones. These features contribute to their lightness and flying ability. The disadvantage is extreme susceptibility to respiratory infections caused by a wide range of bacteria, viruses and fungi. These infections become more of a problem in domestication, which usually involves some degree of increase in stock density - even if only for overnight accommodation - and thus increases the risk of cross-infection. Inadequate ventilation of poultry houses results in a build-up of ammonia gas from poultry faeces, which contain urea. This can predispose the poultry to respiratory disorders, such as sneezing, running eyes and mucous discharges from the mouth. Providing good ventilation easily prevents this.

Common Diseases

The common diseases and disorders of free-range poultry may be either infectious or non-infectious, and are caused by a wide range of organisms or deficiencies. These are summarised in Table 3

Table 3: Causes and examples of poultry diseases.

Causal Agent	***Example***
Infectious	
Virus	Newcastle Disease, Avian Encephalomyelitis, Fowl Pox, Marek's Disease, Infectious Bronchitis Infectious Laryngotracheitis, Gumboro Disease (Infectious Bursal Disease), Duck Virus Hepatitis
Mycoplasma	Chronic Respiratory Disease
Bacteria	Fowl Cholera, Salmonellosis, Pullorum, Fowl Typhoid, Infectious Sinusitis, Colibacilosis
Parasites	*Ectoparasites:* lice, mites, ticks *Endoparasites:* nematodes, Histomoniasis, Haemoparasites, round worms, hair worms, Avian Malaria *Protozoa:* Coccidiosis, Blackhead
Fungus	Aspergillosis: *A. flavis* (toxins), *A. fumigatus* (airsaculitis)
Non-Infectious	
Deficiencies	rickets, curled toe paralysis, encephalomalacia
Toxicities	salt poisoning, food poisoning (Botulism *Clostridium botulinum* and *C. perfringens*), poisonous plants

Infectious Diseases

Viral Diseases

Viral diseases are some of the most important infectious diseases affecting poultry. They are characterised by not being able to be treated, but most can be prevented with vaccines. The more important viral diseases are outlined below.

Newcastle Disease (ND)

This disease (called Ranikhet Disease in Asia) spreads rapidly via airborne droplets spread by the coughing or sneezing of infected birds. The virus can be carried by wild birds, through contaminated eggs, and on clothing. As mortality is often 100 percent in young chickens, ND is probably the most important constraint to family poultry development. Birds of any age can be affected, although young ones are more susceptible. Mortality in older chickens is usually lower, but egg production is usually severely reduced.

The incubation period of three to five days is followed by dullness, coughing, sneezing and gasping. Rapid breathing is accompanied by a gurgling noise in the throat. The respiratory signs usually develop first and are sometimes followed by nervous signs, characterized by twisting of the neck, sometimes combined with dragging of wings and legs. Depending on the environment and the degree of resistance of the birds, not all symptoms may be shown, or they may be in a mild or subclinical form. Some farmers have observed that the twisting of the neck occurs only in birds that survive. Early loss of appetite results in a greenish diarrhoea. The high incidence of ND among family free-range flocks is due to the following factors:

- the prevalence of virulent strains in tropical countries;
- continuous contact with other domestic and wild species of birds (such as ducks and pigeons), which can carry the virus without showing the disease; and
- uncontrolled movement of birds between villages.

There is a seasonal pattern to outbreaks of ND, influenced by:

- the arrival of migratory birds;
- changes in climatic conditions leading to stress, which predisposes birds to the disease;
- hot, dry and windy periods, which encourage airborne spread of the virus; and

- overuse of the few supply points of water available (during the dry season), which then become heavily contaminated with the virus.

Fowl pox

Fowl pox is still prevalent in many poultry flocks, for the following reasons:

- The fowl pox virus can remain alive in the pox scabs (which have fallen off the birds) for up to ten years, which contaminate the environment.
- Mosquitoes and other blood-sucking insects can transmit the virus.

The disease tends to be seasonal, occurring after mosquito breeding times. It is endemic in Papua New Guinea, where it is significant economically because the only NDV in the country is the non-symptomatic form. It is also a major disease in many other tropical countries.

Marek's Disease

Infection occurs early in life, and once a bird is infected, it can shed the virus in skin flakes throughout its life, if it survives. Clinical signs occur in young growing birds in the Acute Marek's Disease (MD) form, characterised by high mortality from visceral tumours. Another peak of mortality occurs in the Classical MD form, characterized by nerve paralysis in the legs and wings of birds aged from 15 weeks to early in the laying period.

Mycoplasmal Diseases

Mycoplasmas are not classified as bacteria or viruses, but as Pleuro-pneumonia-cocci-like organisms (PPLO). These are primarily associated with Chronic Respiratory Disease (CRD), a complex syndrome caused by *Mycoplasma gallisepticum* in partnership with bacteria (often *E. coli*), fungi and viruses (often Infectious Bronchitis). *M. gallisepticum* can be transmitted through the egg. Multi-age flocks, nutritional deficiency and water deprivation are important factors in the epidemiology of the disease in rural poultry flocks.

Bacterial Diseases

Fowl Cholera (Avian Pasteurellosis)

This is a contagious septicaemia (caused by *Pasteurella multocida*) that affects all types of fowls. It is often transmitted by wild birds or other domestic birds, and spreads by contamination of the feed or water and by oral or nasal discharges from infected birds. The incubation period is four to nine days, but acute outbreaks can occur within two days of infection. In some cases, birds die within a few hours of showing the first signs, which vary depending on the form of the disease. The respiratory form is characterized by gasping, coughing and sneezing, while in the septicaemic form there is diarrhoea with wet grey, yellow, or green droppings. In the localized form, the signs are lameness and swelling of legs or wing joints. In acute cases, the head and comb change colour to dark red or purple. If the infection is localized in the region of the ears, a twisted neck (*torticolis*) can sometimes be observed. In chronic cases, the comb is usually pale, with swellings around the eyes and a discharge from the beak or nostril.

Pullorum (Bacillary White Diarrhoea)

This is an egg-transmitted disease (caused by *Salmonella pullorum*) that spreads during incubation or just after hatching. White diarrhoea can be seen from three days to several weeks of age. The chicks refuse to eat, keep their heads tucked in and their wings hanging down. They huddle together and make a peeping sound. Mortality in the acute form ranges from 20 to 80 percent, and in the chronic form is around five percent. In the chronic form, the signs are a marked swelling of the hock joints, poor feather development, lack of appetite and depression.

Fowl Typhoid

Fowl typhoid is caused by *Salmonella gallinarum*, and commonly affects adult fowls. When it occurs in young birds, the signs are similar to those of*S. pullorum*. The incubation period is four to five days, and two days later the birds become depressed and anorexic. The colour of the comb and wattles becomes dark red; the droppings become yellow and the birds close their eyes and keep their heads down. Usually the affected chickens die within three to six days. Pullorum and fowl typhoid complex are both prevalent under free-range conditions.

Avian Salmonellosis (Paratyphoid)

Salmonellosis is usually used to describe infection with any organism of the Salmonella group other than *S. pullorum* or *S. gallinarium*. In countries with intensive poultry systems, poultry meat and eggs are a major source of infection for humans. The opposite may be true of family poultry, with humans infecting poultry. Ojeniyi reported that *S. hirschfeldii* was isolated from cloacal swab samples in fowls and from an adult human male in the same village.

Parasitic Diseases

External Parasites (ectoparasites)

These are very common in scavenging poultry, and include:

- *Lice:* Lice species commonly found on poultry are *Menacanthus straminens, Lipeurus caponis, Monopon gallinae, Goniodes gigas* and*Chelopistes meleagride.*
- *Mites:*Mites such as *Dermanyssus gallinae* can also transmit the bacteria *Borrelia,* which causes fever, depression, cyanosis and anaemia (spirochaetosis).
- *Ticks:* A heavy infestation can produce severe anaemia and, in extreme cases, death due to blood loss. *Argas persicus* is particularly dangerous, being the vector of several blood parasites such as the haemoprotozoa and microfilaria. .

Internal Parasites (endoparasites)

The more important internal parasites are:

- *Helminths (worms):* These are common in scavenging poultry, especially nematodes and cestodes.
- *Protozoa:* The most pathogenic are the coccidiosis disease species of *Eimeria tenella* and *E. necatrix.* Coccidiosis is a common parasitic infection in scavenging poultry.

Fungal Diseases

Mycotoxicosis

The fungus *Aspergillus flavus* commonly grows on stored feed ingredients where moisture content is over eleven percent, especially cereal grains (such as maize [corn]) and oilcake meal (such as groundnut [peanut] meal). The aflatoxin called mycotoxin is the specific toxin produced by *A. flavus.* The toxin itself may remain after all sign of the fungus mould is gone. Ducks are more vulnerable to the toxin (with a lethal dose in the feed of one part per million [ppm] of aflatoxin) than chicken, which can tolerate up to four ppm. In acute forms of the disease, mortality can be as high as 50 percent.

Aspergillosis

This disease is also called airsaculitis. The fungus *Aspergillus fumigatus* causes the disease by growing as a fungus in the lungs and interconnected air sacs. The fungus grows on damp litter or feed, and the bird breathes in the spores, which grow into easily visible lesions as green and yellow nodules, which can completely fill the lungs.

Non-infectious Diseases

Deficiencies

Poultry health is also affected by nutritional and environmental factors, such as insufficient feed or feed deficiencies. A high mortality rate among chicks during the first days or weeks after hatching may be caused by insufficient feed and water. A high mortality in adult birds may be due to nutritional problems, such as salt deficiency. Energy and protein deficiencies and imbalances can arise when the feed contains insufficient quantities of these nutrients, resulting in poor growth in young stock and a drop in egg production and egg weight in laying hens.

Toxicities

An excess of certain nutrients, especially minerals, can cause abnormalities. An excess of common salt (NaCl), for example, results in deformed eggshells as well as increased water consumption, and if drinking water is restricted (as is often the case with free-ranging birds), signs of toxicity may develop. Free access to feed of high carbohydrate and low fat, combined with lack of exercise, high temperatures and stress, can cause Fatty Liver Syndrome, which can result in high mortality.

Disease Control in Family Flocks

Non-medical Disease Control

The most economical and effective means of preventing non-viral diseases is improved management and nutrition, of which the most important aspects are hygiene, housing, flock structure, and young chick care and feeding.

Medical Disease Control

Simple medical control measures appropriate for free-range village flocks include:

- Vaccination against Newcastle Disease, Fowl Pox and Fowl Cholera.
- Deworming for internal parasites in a mixed flock, with a polyvalent poultry dewormer such as Piperazine (added to drinking water). With guinea fowl, a dewormer against *Trichomonas* should be used.

- Treatment for external parasites. Insects and other external parasites build up quickly in poultry huts, coops and baskets. There are effective traditional methods against ectoparasites. All the surfaces of the basket, coop or hut can be sprayed with a suitable insecticide, using the same type of hand-pump used for spraying mosquitoes. This procedure should only be carried out when the house is empty in the morning, and the birds should not be allowed back inside until evening. External parasites living on poultry can best be treated by adding powdered mothballs (naphthalene) and ash to the dust bath area. Ash dust is more abrasive than ordinary soil dust, and thus removes the waxy coating of the insect exoskeleton when the bird takes a dust bath. If enough of the waxy coating is removed, the insect will dehydrate and die.

References

Adeyanju, S.A., Ilori J.O. & Osuntogun, C.A. *Poultry Farming in Ondo State*, Monograph, Ile-Ife, Nigeria. Department of Animal Science, Obafemi Awolowo University. 20 pp.

Baksh, I. (1994). Permaculture for Poultry. *African Farming,* (Jan- Feb), 13-14.

Koster, E. (Ed.) (1996). Poultry Farming Systems for Small Commercial Farmers. *Proceedings of a Poultry Workshop held at the Animal Improvement Institute*, Irene, South Africa.

McInerney, J. (1992). Economic aspects of disease in poultry production. *Proceedings XIX World Poultry Congress*, 2: 561-566.

Odi, H. Diambra. (1990). State of smallholder rural poultry production in Cote d'Ivoire. *Proceedings International Seminar on Smallholder Rural Poultry Production,* 9-13 October, Thessaloniki, Greece, pp.105-116.

ter Horst, K. (1986). *Poultry Management in rural areas of developing countries: Matching technical demands with social facts.* Unpublished report, Department of Livestock Services, Ministry of Livestock and Fisheries, Bangladesh.

8

Milking and Milk Production Hygiene

Humans first learned to regularly consume the milk of other mammals following the domestication of animals during the Neolithic Revolution or the invention of agriculture. This development occurred independently in several places around the world from as early as 9000–7000 BC in Southwest Asia to 3500–3000 BC in the Americas. The most important dairy animals—cattle, sheep and goats—were first domesticated in Southwest Asia, although domestic cattle has been independently derived from wild auroch populations several times since. Initially animals were kept for meat, and archaeologist Andrew Sherratt has suggested that dairying, along with the exploitation of domestic animals for hair and labor, began much later in a separate secondary products revolution in the 4th millennium BC. Sherratt's model is not supported by recent findings, based on the analysis of lipid residue in prehistoric pottery, that show that dairying was practiced in the early phases of agriculture in Southwest Asia, by at least the 7th millennium BC.

From Southwest Asia domestic dairy animals spread to Europe (beginning around 7000 BC but not reaching Britain and Scandinavia until after 4000 BC), and South Asia (7000–5500 BC). The first farmers in central Europe and Britain milked their animals. Pastoral and pastoral nomadic economies, which rely predominantly or exclusively on domestic animals and their products rather than crop farming, were developed as European farmers moved into the Pontic-Caspian steppe in the 4th millennium BC, and subsequently spread across much of the Eurasian steppe. Sheep and goats were introduced to Africa from Southwest Asia, but African cattle may have been independently domesticated around 7000–6000 BC. Camels, domesticated in central Arabia in the 4th millennium BC, have also been used as a dairy animal in North Africa and the Arabian peninsula. In the rest of the world (i.e., East and Southeast Asia, the Americas and Australia) milk and dairy products were historically not a large part of the diet, either because they remained populated by hunter-gatherers who did not keep animals or the local agricultural economies did not include domesticated dairy species. Milk consumption became common in these regions comparatively recently, as a consequence of European colonialism and political domination over much of the world in the last 500 years.

All female mammals can by definition produce milk, but cow milk dominates commercial production. In 2011, FAO estimates 85% of all milk worldwide was produced from cows. In the Western world, cow's milk is produced on an industrial scale and is by far the most commonly consumed form of milk. Commercial dairy farming using automated milking equipment produces the vast majority of milk in developed countries. Aside from cattle, many kinds of livestock provide milk used by humans for dairy products. These animals include buffalo, goat, sheep, camel, donkey, horse, reindeer, and yak.

In 2010, the largest producer of milk and milk products was India followed by the United States, China, Germany, Pakistan and Russia. The 27 countries of the European Union together produced about 138 million tonnes of milk in 2011. Increasing affluence in developing countries, as well as increased promotion of milk and milk products, has led to a rise in milk consumption in developing countries in recent years. In turn, the opportunities presented by these growing markets have attracted investment by multinational dairy firms. Nevertheless, in many countries production remains on a small scale and presents significant opportunities for diversification of income sources by small farmers. Local milk collection centers, where milk is collected and chilled prior to being transferred to urban dairies, are a good example of where farmers have been able to work on a cooperative basis, particularly in countries such as India.

A survey over 2001 and 2007 was conducted by ICAR (International Committee for Animal Recording) across 17 developed countries. The survey found that the average herd size in these developed countries increased from 74 to 99 cows per herd between 2001 to 2007. A dairy farm had an average of 19 cows per herd in Norway, and 337 in New Zealand.

Annual milk production in the same period increased from 7,726 to 8,550 kg per cow in these developed countries. The lowest average production was in New Zealand at 3,974 kg per cow. The milk yield per cow depended on production systems, nutrition of the cows, and only to a minor extent different genetic potential of the animals. What the cow ate made the most impact on the production obtained. New Zealand cows with the lowest yield per year grazed all year, in contrast to Israel with the highest yield where the cows ate in barns with energy-rich mixed diet.

Basics of Milking

Dairy farmers, with varying levels of skill, knowledge and resources, maximise returns from milk production by influencing lactation through selective breeding and control of reproduction, nutrition, disease and general management. The methods of milking have a particularly important effect because a cow cannot secrete over a period more milk than is removed by milking. Thus, maximising milk removal in ways which are economic will take fullest advantage of secretion potential.

Lactation

Lactation includes both milk secretion and storage in alveolar cells and ducts within the mammary gland, followed by milk ejection (let-down) and milk removal. Milk secretion is continuous and usually at a constant rate for at least 12 hours resulting in a gradual increase in internal udder pressure. Milk ejection is a neuro-hormonal reflex initiated by various stimuli at milking time. These stimuli, which reflect good husbandry practices, are either natural (inborn) or conditioned (learned by experience), including, for example, feeding and udder preparation. They cause the alveoli and small milk ducts to contract forcing milk towards the udder sinus. Once this has happened most, but not all, of the milk can be removed when external forces such as suckling or milking open the streak canal (teat duct) at the teat end, but at least 10% will be retained in the udder as residual milk.

Milking Intervals

The 10%–20% of the secreted milk which is not expressed from the secretary tissue and is retained in the udder when milking is completed is called 'residual milk' and has a much higher fat content than even the end-of-milking strippings. The quantity of residual milk is proportional to total yield, so that with unequal milking intervals there is a larger net carryover of milk fat from the longer night-time to the shorter daytime interval. This accounts for the apparent faster secretion rate and higher fat content of afternoon milk production. Milk yields, particularly from higher yielding cows are usually greater when milking intervals are 12 hourly. The effect of uneven intervals is not large up to 16 and 8 hours, and can be minimised by milking the higher yielders first in the morning and last in the afternoon.

Milking Frequency

As a general rule, herd lactation yields will rise as the frequency of milking is increased. On average, the rise in milk yields will be between 10% and 15%, the largest increases occuring amongst heifers. The chemical composition of the milk (fat and solids-not-fat) will be unaffected. Recent commercial data from developed dairying areas also reveal that, on average, up to 10% increase in yield is required to cover the extra costs of milking three times daily. The full benefit of the increased frequency is obtained by milking three times daily throughout lactation rather than reverting to twice daily when milk yields begin to fall. The reasons for the increase in lactation yields are inconclusive; the most likely being the more frequent removal of secretion inhibiting substances which begin the drying-off process.

Incomplete Milking

There are two forms of incomplete milking. One is that excessive amounts of residual milk are retained in the udder because of inadequate milk ejection stimuli or the inhibitory effects of adrenalin secreted by cows becoming frightened and upset during milking, or even by slow milk removal. The other form is that some of the available milk is left in the udder when milking ceases, ie., the so-called 'strippings'. The modern milking machine is designed to remove 95% of available milk without recourse to additional cluster weight or manual assistance. Hand stripping, particularly with the finger and thumb should be avoided. The amounts of strippings are likely to be small even in relation to normal levels of residual milk and if not removed are unlikely to affect significantly either the lactation yield or quality of milk.

Milking Routines

The aim of an efficient and effective milking routine is to leave the least amount of residual milk in the udder. This, in itself, is a measure of good stockmanship. Milk ejection can be stimulated manually by a series of activities carried out by the person doing the milking. The amount of residual milk is inversely proportional to the strength of the conditioned stimuli signals, which are developed into a regular, repetitive milking routine, including such activities as feeding and udder preparation. The stimulation response is transitory, the maximum effect declining within a few minutes of milk ejection occuring. Therefore delayed milking will reduce the amount of milk removed. The internal pressure of milk within the udder peaks between one and two minutes after milk ejection and therefore milking should be completed as soon as possible after this occurs. Cows are creatures of habit and consequently changes to the routine should be made gently and quietly. It is important to avoid any circumstances which upset or frighten them causing the release of adrenalin which adversely affects the circulatory and musculatory systems, thus restricting effective milk ejection and prolonging the duration of milking. The response of cows and those milking them to a pleasant and stress free environment will be measured in terms of production levels.

Because residual milk and strippings have fat percentages that normally exceed 10%, incomplete or slow milking can reduce markedly the fat content of the milk at any particular milking. However, it is important to recognise that milk fat retained or left in the udder is not lost but will be obtained at succeeding milkings. In fact, although management factors (eg. varying milking intervals and milking frequency) may alter the fat content of milk at one milking, the average fat content over a period of time will be unaffected. On average, the fat content of milk obtained must be the same as that secreted into the udder. The concentrations of protein, lactose and other solids-not-fat are unaffected by changes in milking management either at one or more milkings.

Systems of Milking

Hand Milking

Statistics from all major milk producing countries indicate an annual decline in the number and size of hand milked herds. Labour productivity in these herds is low with very few cows per person involved. Duration of milking each cow and the whole herd is protracted because each person milks cows one at a time with a relatively slow milk extraction rate compared with machine milking. These are factors which contribute to lower average lactation milk yields in hand milked herds. Nevertheless, for small herds, hand milking will usually be the method chosen because where there is sufficient labour available it can provide a satisfactory way of milk removal with minimal capital investment, equipment maintenance and cleaning.

During milking, hygiene standards require clean milking clothes and hooded milking buckets to prevent dust, dirt and udder hairs falling into the milk. Udders and tails need regular clipping. Before milking begins the foremilk is drawn and examined and visible dirt removed from udders and teats by washing and drying with disposable paper towels. Milk with clean, dry hands using the full hand in preference to just finger and thumb, a practice which can lead to misshapen udders and teat injury. It is best to milk rear quarters first as they contain the higher proportion of milk.

Milk cooling methods will depend mainly on the local water supply. If the quality, quantity or temperature are unsuitable or unreliable, then the milk should be taken, within 3 hours of production, to a central depot for cooling. Where an unlimited free supply of clean, cold (below 15°C) water is available, the cans of milk can be immersed in running water. Water usage can be reduced and cooling rate increased by inserting a turbine in-can cooler into the cans of milk. Alternatively, the milk may be tipped and allowed to flow over a corrugated surface cooler connected to the water supply.

Cleaning the milking bucket and cooler is best done by an initial rinse in clean water immediately after milking, followed by scrubbing in a hot (45°C or above) detergent/disinfectant solution before finally rinsing in chlorinated water (50 ppm). Alternatively, after scrubbing the equipment in hot detergent solution, disinfect by immersing it in hot (above 75°C) water for at least 3 minutes. The foremilk cup, stool and udder washing equipment should be treated similarly afterwards. All equipment must be drained dry during the interval between milkings.

Bucket Machine Milking

Bucket milking machines were the first major development in the mechanisation of milking systems and were designed particularly for herds kept in cowsheds. Each portable unit, consisting of a 15 litre capacity lidded bucket, pulsator and teat-cup assembly or cluster, requires manual attachment to a vacuum supply when it is moved from cow to cow during milking. Milk is tipped from the buckets into milk cans positioned in the dairy (milk room) or in the cowshed. The system is mechanically simple with relatively low investment, running and maintenance costs compared with milking machines in parlours. Milking performance is restricted in terms of cows milked per hour by the amount of work that must be carried out on each cow (the work routine). A high proportion of time and effort is spent walking from cow to cow and manually carrying equipment and transporting, lifting and tipping milk. As a result, each operator is unlikely to be able to use effectively more than 2 or 3 bucket machine units and, consequently, will not milk more than about 30 cows per hour. A better performance will be achieved by using a trolley carrying the milk cans, spare milking machine bucket and udder preparation equipment. The trolley is moved along the cowshed as milking proceeds. The consequent reduction in time and effort will allow an additional unit to be used thus raising potential performance to about 40 cows per hour.

Milk tipped direct to milk cans during milking is cooled by inserting an in-can turbine cooler when the cans are full. Otherwise, it will be necessary to carry the buckets of milk into the dairy and pour the milk over a corrugated surface cooler coupled to the water supply. Both methods are equally effective although using a surface cooler is more laborious. In either case, cooling efficiency is improved and water requirements reduced by connecting the cooler to a chilled water unit.

Cleaning bucket milking machines and the ancillary equipment is a three-part manual operation using detergent/disinfectant solutions. Immediately after milking, visible dirt and milk deposits on the outside of the equipment are removed with clean, cold water. Each unit is then connected to the vacuum supply and 10 litres of clean water drawn through the teat-cups into the unit bucket. This initial rinsing is followed by scrubbing all the milking equipment (except pulsators and pulse tubes) in a hot (not less than 45°C) detergent/disinfectant solution using 45 litres per 3 milking units. Volumes of water and quantities of chemicals used must be carefully measured. Finally, the equipment is rinsed in chlorinated water (50 ppm) and allowed to drain dry in a clean place. Ancillary equipment is treated similarly afterwards.

Direct-to-Can Milking

About 1950, many herds increased in size and moved from cowsheds to abreast parlour milking. The direct-to-can milking system was devised to meet this change and, consequently, it is applicable to this particular parlour. Compared with bucket machines, the capital investment, running costs and labour input are lower. As a complete system of milking, cooling and cleaning it is very simple indeed.

Milk is drawn direct from the udders to the milk cans via a specially designed lid which connects the milk can to the vacuum supply. Manual lifting, carrying and tipping of milk is eliminated with a consequent improvement in milking efficiency and performance. Therefore, each operator is able to manage a 4 unit, 8 stall or a 5 unit, 10 stall abreast milking parlour effectively. When full, the milk can is replaced with an empty one and milk cooling may begin. Cooling milk produced with the direct-to-can system is done simply by inserting an in-can cooler into the cans of milk and connecting it either to a clean, cold water supply or to a chilled water unit.

Cleaning direct-to-can milking equipment also provides substantial economies in labour and other costs of cleaning using a method developed specifically for the purpose. In addition to being bactericidal the cleaning treatment prevents milk fat accumulation in natural rubber components. When milking is finished, all visible dirt and milk deposits are removed from the milking equipment by rinsing in clean water. Next, the component parts of each unit, namely, the stainless steel lid and rubber gasket, two long rubber tubes and a stainless steel teatcup assembly are loaded into a purpose designed steel-mesh basket (up to a maximum of 3 units). The parts are positioned in the basket in such a way as to avoid any airlocks when lowered into a rubber or mild-steel bin containing a 3% to 5% caustic soda solution. A rubber lid covers the bin and the equipment remains in the solution until the next milking when it is lifted out to allow surplus solution to drain back into the bin. Before use, the equipment is removed from the bin and all traces of caustic soda solution removed by thorough rinsing in chlorinated (50 ppm) water. At monthly intervals the solution is renewed and ethylene diamine tetra-acetic acid (EDTA) added at the rate of 60gms to 45 litres of solution to prevent hard water deposits. After each milking the cooler and ancillary equipment are cleaned in the way described for bucket milking machine

Pipeline Milking

The introduction of bulk milk collection and refrigerated milk tanks on farms, together with the development of large static and rotary parlours for milking big herds, gave an impetus to pipeline milking systems which hitherto had been installed in large cowsheds and milking barns. The main advantages are that the milk is transported under vacuum from udder to dairy for cooling and storage and the cleaning and disinfection of the milking equipment can be done in-situ with very little manual involvement. In addition, devices can be inserted into the milking pipeline to reveal clinical signs of mastitis, indicate the milk yield from each cow, allow samples to be taken and automatically remove the cluster when milk flow ceases (thus eliminating overmilking). Internationally agreed standards prescribe the minimal diameter of pipelines to enable the milk to bc transported without adversely affecting vacuum stability at the cluster. These comparatively high investment, low labour cost systems are the only practical alternative for large and medium sized herds milked in parlours, particularly where bulk milk collection is involved. During milking, operator work routines can be reduced to assisting cow entry and exit, udder preparation and cluster attachment so that milking performances of more than 85 cows per manhour can be achieved.

Milk cooling can be done by discharging the milk over a corrugated surface cooler connected to the water supply or a chilled water unit and collecting it in milk cans underneath. Alternatively, the milk can be pumped direct from a milk receiving vessel to a refrigerated bulk milk tank or via a pre-cooler to an insulated milk storage tank.

CLEAN and disinfect the pipeline milking plant in-situ by first removing manually any visible dirt and milk deposits from external surfaces and making the necessary adjustments to form a complete circuit between milking and milk transfer pipelines. Recirculate hot detergent/disinfection solution when the initial hot water rinse reaches 65°C at the discharge point, for 10–15 minutes at 10–15 litres per unit. Finally rinse with chlorinated water at 50 ppm.

Alternatively, use a single, once-through circulation of near boiling water at 15–20 litres per unit, and maintain a temperature of not less than 77°C FOR AT LEAST 2 minutes. During the first 2–3 minutes of flow, entrain 1 litre of diluted nitric or sulphamic acid.

Cleaning and disinfecting pipeline milking equipment in-situ can be manually controlled or automatic using sequence timers which initiate the selected programme. Both the two available methods begin with the manual removal of visible dirt and milk deposits. Simultaneously, adjustments are made to enable a complete circuit to be formed incorporating the milking and milk transfer lines. Recirculation cleaning is a three-stage process incorporating a hot water pre-rinse which is discharged to waste until the water at the discharge point reaches 65°C. Detergent and disinfectant is then added to the hot water and the solution re-circulated (at 10–15 litres per unit) for 10– 15 minutes. Finally, chlorinated water is circulated once and discharged to waste. Acidified Boiling Water (ABW) cleaning relies solely on heat disinfection. It consists of a single, once-through circulation of near boiling water (at 15–20 litres per unit), maintaining a temperature above 77°C on all milk contact surfaces for at least 2 minutes. During the first 2–3 minutes of flow 1 litre of dilute acid (70% w/w nitric acid or sulphamic acid crystals) is drawn into the boiling water to prevent deposition of hard water salts. Simple modifications are inserted into the pipeline circuitory to ensure a balanced flow of the hot solution at the desired rate to all milk contact surfaces. The refrigerated bulk milk tank is cleaned manually or mechanically.

Milking Machines and Equipments

Methods of machine milking are designed to create a pleasant milking sensation for the cows and to avoid any possible hazard to udder health. It is most important that milking is done with a well designed, carefully cleaned and properly maintained machine which is operated strictly according to the manufacturer's instructions. A skilled operator pays particular attention to careful cluster attachment and removal from the udder. During cluster attachment it is essential to ensure that the vacuum cut-off arrangements to the clawpiece are effective so that excessive volumes of air do not enter and cause vacuum fluctuations in the main vacuum pipeline system, as this could increase mastitis incidence. Attach each teatcup carefully starting with the two furthest from the operator. The clusters are removed as soon as milk flow ceases, avoiding excessive air entry through the teatcups by cutting off the vacuum supply before gently but firmly pulling the teatcups from the udder. During milking, any teatcups which slip from the teats should be readjusted immediately and any clusters which fall to the floor should be cleaned and reattached without delay.

Basic Construction

The *principle* of machine milking is to extract milk from the cow by vacuum. The machines are designed to apply a constant vacuum to the end of the teat to suck the milk out and convey it to a suitable container, and to give a periodic squeeze applied externally to the whole of the teat to maintain blood circulation.

A milking machine installation consists of a pipework system linking various vessels and other components which together provide the flow paths for air and milk. The forces necessary to move air and milk through the system arise from the fact that it is maintained at a vacuum. Thus it is atmospheric pressure which forces air, and intra-mammary milk pressure which forces milk, into the system and the combination of these forces causes flow. To be a continuous operation it is necessary to remove air and milk from the system at appropriate rates.

Although milking machines have now developed into systems that show considerable diversity they have the same basic components. The air is removed by *a vacuum pump* at a constant rate. In a bucket or direct-to-can machine milk is removed from the system by disconnecting the milk container; in milking *pipelineand recorder* machines the milk is removed by a *milk pump or releaser.*

Vacuum and Milk Flow

When the milk from the claw is raised to a pipeline this can markedly reduce the vacuum at the teat because of the weight of milk in the long milk tube. The reduction in vacuum can be much reduced by bleeding air through a small hole in the clawpiece

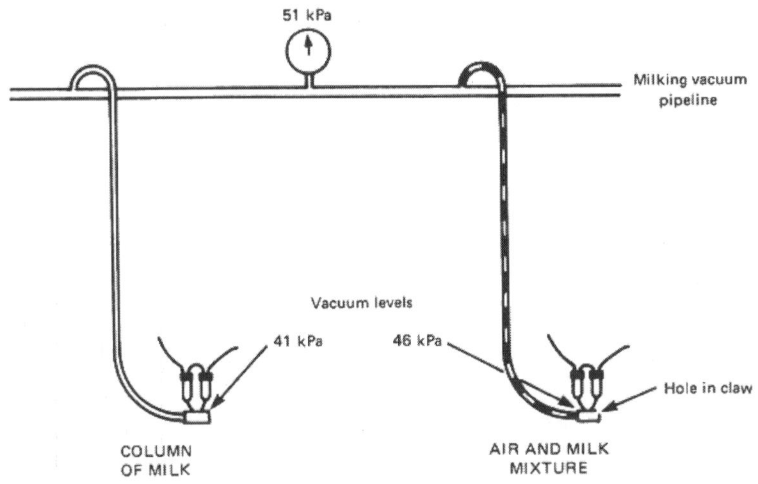

In addition to the designed sources of air admission, air can be drawn into the teatcups past the teat and also when a milk container is changed or

emptied. In a poorly maintained machine there may also be inward leakage of air at joints or points of damage. To maintain the working vacuum the vacuum pump extracts the air admitted into the system by compressing it so that it can be discharged to atmosphere.

In pipeline milking machines the flow pattern is similar to the bucket machine except that milk and air from each claw flow either directly to a *recorder* vessel where air and milk are separated, and/or through the milking pipeline to a common receiver vessel where milk and air are separated. There is no further air admission at this point when a motor driven releaser *milk pump* is used to empty the receiver. Other types of releaser (eg. pulsator controlled spit chamber and double chambered weight operated) admit air. Where air and milk are transported together the flow pattern becomes complex depending on various factors particularly the volume of air relative to milk or air:milk ratio. Air is normally admitted in to the claw at a rate of 4 to 8 l/min. A milk flowrate for a fast milking cow will be about 6 l/min, giving an air:milk ratio of 0.7:1 to 1.2:1. Towards the end of milking when the milk flowrate has decreased to 0.25 l/min the ratio becomes 16:1 to 32:1.

The air:milk ratio becomes important where milk has to be elevated from the claw as in milk pipeline and recorder machines other than those with low level milk pipelines. Elevating a liquid, as distinct from a gas, involves a loss of potential energy and this is compensated for by a change of vacuum. Thus elevating a column of milk in a vacuum system through 1 m height reduces the vacuum by about 10 kpa. Therefore if the vacuum at the top of the column is 51 kpa it will be only 41 kpa t the bottom. This vacuum drop is markedly reduced by the admixture of air. If the air:milk ratio is 1:1 the weight of milk in the column is halved and the vacuum drop becomes only 5 kpa; if it is 9:1 the vacuum drop is only 1 kpa. Under vacuum liquids cannot flow against gravity except as a column which fills the bore of the tube. Where the tube contains air and milk the liquid forms plugs which are separated by pockets of air in the proportion determined by the air:milk ratio.

Measurement of Vacuum

Vacuum is a pressure below atmospheric pressure, the term "negative pressure" is sometimes used but in milking machine terms it may be considered to mean "vacuum" measured on a scale in which atmospheric pressure at the time and place of measurement is zero vacuum.

Vacuum can be measured in a variety of units. A commonly used measure is the linear difference in height between two columns of mercury in a 'U' tube when one of the columns of mercury is subjected to a vacuum and the other open to atmosphere. The difference in height of the levels is supported by atmospheric pressure.

In the past the most commonly used units have been inches, millimetres or centimetres of mercury (in Hg, mmHg or cmHg). Units now adopted by the International Standards Organisation (ISO) for International Standards of milking machines for vacuum measurement are kilopascals (kPa) with zero (0) kPa being equal to atmospheric pressure and 100 kPa absolute vacuum. Equivalent relationships for values of vacuum levels are:-

1 mmHg = 0.133 kPa

1 inHg = 3.386 kPa

Equivalents for vacuum levels of 50 kPa and 44 kPa that are the most commonly used levels for milking cows:

50 kPa = 14.8 inHg = 375 mmHg

44 kPa = 13.0 inHg = 330 mmHg

Action of the Milking Machine

The principles of machine milking were established many years ago and the basic method described below, is used in virtually all commercial milking machines although in a minority some modifications are made. The teatcup liner is the only equipment that comes into contact with the cows teats. The continuous vacuum within the liner causes the teat duct (streak canal) to open and the milk to flow because of the pressure difference between the milk in the teat and vacuum. To prevent damage or pain to the teat that would be caused by the continuous vacuum a system called pulsation is used. This makes the liners collapse on and below the teats about once each second massaging the teat and maintaining a more normal blood flow. In each pulsation cycle milk does not flow from the teat when the collapsed liner squeezes the teat duct.

Providing the cows 'let down' (ejection) has occurred the flow rate from the teat depends largely on the bore of the teat duct which is an inherent factor and not subject to management practices or training. Flow rates are also influenced by the mechanical properties of the milking machine. After the teat cups have been attached the flow rate reaches a maximum in about one minute, usually within the range of 2–5 kg/minute and the total milk flow period will range from 2 to about 8 minutes depending upon milk yield. Flow rates decline at the end of milking and when flow ceases there is usually a small amount of milk trapped in the sinus of the udder which can be removed by pulling ownwards on the clawpiece and massaging the udder (ie. machine stripping). With modern designs of liner the quantity of strippings is small (ie. less than 0.3 kg) and machine stripping is not usually practiced. The small amounts of milk that are left do not affect milk yield or the average chemical composition of the milk obtained or mastitis.

Components

Vacuum System

The function of the *vacuum pump* is to extract air from the pipeline system and in the majority of milking machines it is a rotary exhauster driven from either an electric motor or a stationary internal combustion engine, using pulleys and V-belts. This gives flexibility with speeds ranging from about 800–1500 rev/min and a corresponding ange of pump capacities to suit several sizes of milking machines. A few vacuum pumps are direct shaft coupled; their speed is then motor speed and cannot be changed. When operating with a vacuum at the pump inlet suitable for milking cows (50 kPa), the pumping capacity should be sufficient for the total number of milking units connected. This will range from 70 litres of air per minute per unit for installations with 20 units to 85 per litres per minute for the smaller plants with 5 units. It will vary according to individual manufacturers design of equipment and the additional ancillary components included. Oil lubricated high speed vacuum pumps operate at speeds of between 850–1440 revs/min. For each 0.75 kW (1 h.p.) of rated electric motor requirement the extraction capacity is approximately 280 litres of air/min. The pump should be fitted with drive belt guards, an effective silencer and, for oil lubricated pumps, an oil separator. Also, a nonreturn valve should be fitted in the exhaust pipe to prevent reverse rotation of the vacuum pump when switching off at the end of milking. This is to avoid debris from the silencer/ exhaust pipe being sucked back into the pump. To prevent solid and liquid material being drawn into the pump from the milking installation, an *interceptor/trap vessel* with a capacity of not less than 15 litres is fitted in the main vacuum line adjacent to the pump, with provision for draining and cleaning. It should also have a float valve to shut off the vacuum in the event of it filling with liquid.

Cluster

The *cluster* which attaches to the cow, consists of four teatcup assemblies (each having a shell, a rubber liner and a short milk and short pulse tube), a claw, a long milk tube and long pulse tube(s). Teatcup shells are normally made of stainless steel. Plastics or a combination of plastics and metal are also used. The liner is a flexible rubber sleeve having a mouthpiece, and when assembled in the shell under tension, forms an annular space (pulsation chamber) between the liner and shell. This pulsation chamber is connected to the pulsator through a nipple on the side of the shell via the claw. The teatcup assemblies are connected by short milk and short pulse tubes to the claw, which is connected to the milking and pulsation vacuum by a long milk tube and long pulse tube(s). To stabilise the vacuum in the teatcups during milking, the claw has a small air admission hole, about 0.8 mm in diameter, which admits approximately 7–8 litres of air/min into the bowl of the claw. This air helps to carry the milk away, preventing flooding and violent vacuum fluctuations. The claw is made of stainless steel or a combination of plastics and stainless steel, and usually weighs about 0.5 kg and the total all up weight of a milking cluster is about 2.5 kg. The weight of a milking cluster is important and the correct weight relates to the design of liners. Too little weight gives incomplete milking because of high levels of strippings, too much weight will result in milking units falling off during milking.

The bore of the rubber short milk tubes should not be less than 8 mm and the short pulse tubes not less than 5 mm, and the long milk tube should not be less than 12.5 mm. The effective claw bowl volume should not be less than 80 ml.

Milk and Air Separation

In pipeline milking installations it is necessary to include a method of extracting milk from the vacuum system. A *receiver* vessel is fitted to act as a milk reservoir and air separator and from this the milk is pumped out. It is made of either glass or stainless steel and may have a capacity of 35 to 160 litres or more depending on the method of cooling and storing the milk. During milking the weight or level of milk in the receiver is used to start the milk pump (releaser).*Releaser milk pumps* for extracting milk out of the vacuum system are: centrifugal, with capacities of at least 4550 litres per hour, or diaphragm, with capacities of approximately 2000 litres/hr for a single ended pump. Other methods for extracting milk from a vacuum system are double chambered releasers with weight operated mechanical valves and the 'spit chamber' releaser. These do not require electricity and the latter utilises a pulsating vacuum to alternately open and close flap valves allowing the milk to drain out of the second chamber.

In addition to the *interceptor* for the vacuum pump, a *sanitary trap*, is fitted in the vacuum pipeline adjacent to the receiver. This is a glass vessel of not less than 3 litres capacity that separates the part of the milking machine through which milk passes from the air system, preventing movement of liquid from one to the other. If milk enters the sanitary trap it is an indication of a fault in the machine. Therefore, it should be mounted within sight of the milker and be fitted with a float ball to shut off the vacuum. The vacuum connection between the sanitary trap and the receiver should slope away from the receiver to the sanitary trap.

Recording and Sampling

In pipeline recorder milking installations, milk flows directly from the cluster into a rigidly mounted calibrated *glass recorder jar where* the milk is intercepted and held to allow measurement of the total individual cows milk yield. This is done by using either the calibrations etched on the side of the glass jar or the weight of the contents for jars that are supported on electrical strain beams.

In direct to pipeline milking machines not fitted with recorder jars, milk yields are measured with *milk meters*. There are several designs of meters. Basically there are two distinct methods used. (1) Proportional flow; these collect a constant proportion (about 2.5%) of the milk flow into a calibrated flask the contents of which are recorded manually or electrically. (2) Batch; the total flow of milk is continuously measured in discrete batches and the aggregate of the batches recorded.

Proportional flow meters used for manual recording are relatively simple and inexpensive. Electrical recording of milk yields, depending on complexity can be costly and are usually linked with electronic management data recording systems.

All these methods of milk recording have the facility for collecting a milk sample for chemical analysis and are required to be within set limits of accuracy for yield and ability to collect a true representative sample of the milk. These limits are laid down in an International Standard for milk yield recording equipment by I.C.R.P.M.A.

Milk Flow Detection and Cluster Removal

With all milking units, except those using transparent recorder jars, there is some provision made for observing milk flow. This can be a metal claw fitted with "windows", or a claw made in a transparent material or, a short length of clear tubing fitted in the long milk tube. Observing flow in these ways can be misleading and there are now milk flow indicator devices that work on the basis of milk flowing into a small container with a submerged metering orifice for maintaining a milk level. Variations in this level can be observed or detected electrically, using a magnetic float switch or electric probes.

In modern large milking installations, milk flow detectors are used to sense the end of milk flow (less than 250 g/min) and activate automatic cluster removal (A.C.R.) equipment. This equipment prevents overmilking and enables the operator to use more units by reducing the workload. It eliminates waiting to observe when an animal has finished milking and manually detaching the cluster. Electrically operated milk yield recording equipment eg. glass recorder jars on strain beams and milk meters can be used as flow detectors and linked to A.C.R. equipment. In this way the milking cluster is removed when flow rate has fallen below a predetermined level. A useful device that can be fitted in the long milk tube is a small in *line filter* used as a *mastitis detector*. It retains milk clots enabling clinical mastitis to be detected and also acts as a coarse filter for removing extraneous material which is an advantage when using flow detectors or milk meters that have small metering orifices etc. The filter screen is easily removed for rinsing off the clots after each cow and the detector is cleaned in situ by the normal cleaning system of the milking machine.

To reduce or eliminate the effect of vacuum fluctuations that occur during milking "*shields*" can be fitted in the bottom of the liners near the outlet preventing contaminated milk returning up the short milk tube from impinging on the teat end. The shield consists of a disc of stainless steel or plastics of a diameter that leaves a space etween the edge of the disc and the liner wall not less than the internal diameter of the milk outlet. A more positive method to prevent the reverse flow in a conventional cluster is to fit *non-return valves* in the claw or short milk tubes together with a small air admission hole between the liner and the valve to maintain normal milk flow. A system of milking with non-return valves but without the air admission holes has been investigated and preliminary results show the development could have advantages.

Water Heating

For nearly all milking machine cleaning methods an adequate supply of hot water is necessary. *Water heaters* vary from simple free standing insulated or uninsulated tanks filled with a hose and manual control to insulated tanks automatically filled and heated at predetermined times.

The water capacity required is dependant on the washing system used (hand or in-place cleaning) and the number of units to be washed. For in-place cleaning the amount of water required for each milking unit will be 12 literes of water at 85°C for recirculation cleaning and 18 litres at 96°C for a single pass "boiling water" cleaning system. Additional hot/warm water may be required for udder washing and calf feeding. Whatever cleaning methods are used the water heater should be capable of producing near boiling water for the purpose of giving the installation a periodic heat treatment. The temperature of the wash solution in closed pipe work is affected by the vapour pressure of the liquid. Water at normal atmospheric pressure at sea level boils at 100°C (212°F) and at a vacuum of 50 kPa (14.8 inHg) it boils at 81.7°C (179°F). Thus when cleaning pipeline milking machines with water solutions under vacuum, the maximum temperature which can be obtained ill depend on the operating vacuum level in the installation, ie., the higher the vacuum level the lower will be the temperature that can be obtained. For similar reasons, at an altitude of a 1,000 metres where atmospheric pressure is reduced to about 89 kPa (12.9 lb/in^2), water in an open vessel will boil at the lower temperature of 96°C (205°F) instead of 100°C (212°F) at sea level.

Although any fuel can be used for water heating the most commonly used is electricity for cleanliness and convenience. In areas with hard water supplies, a hard scale will build up in the tank and on the heating element. The electrical heating element should be fitted with a clear space beneath it (approx. 10 cm) and be of a type that will flex during heating and cooling, breaking off the hard water deposits that will fall into the space below. Provision should be made for easy removal of this accumulated scale.

Maintenance

The basic layout and operation of milking machines is normally straightforward and similar for all standard milking machines. However, it is important that a correct maintenance routine is followed. Faulty milking machines can result in poor milk let down, slow milking, milk of high bacterial count and mastitis. Maintenance is a responsibility of the person regularly using the equipment with simple checks made at each milking and more detailed ones at weekly or less frequent intervals.

The maintenance instructions described deal with the essential routines and do not include checks on the many optional items not fitted in large modern parlours, (eg. automatic cluster removers, milk meters). These vary considerably and the maintenance procedure should be obtained from the supplier. Because the overall performance can be assessed only with special measuring equipment a fullscale test should be carried out by the manufacturer, installer or extension service at least once per year.

Milk Pumps

The performance of a centrifugal milk pump can be seriously affected, even made inoperative, by air leaks at the rubber seal on the drive shaft or the non-return flap valve fitted in the pump outlet or delivery pipe. Where glass receivers are in use, test for leaks by putting approximately 4 litres of water in the receiver with the milk pump not working. With the receiver vessel under vacuum look for air bubbles rising to the surface from the receiver outlet. If a continuous stream of bubbles is seen dismantle the non-return valve and examine rubber flap for wear and distortion, renew if necessary. If leaks continue replace pump or dismantle and renew shaft seal, inspecting shaft for wear. It will be necessary to replace pump if there are signs of annular grooves on shaft.

In recent years the International Standards Organisation (ISO) a world federation of national standards institutes (ISO member bodies) has worked with other organisations for a standardisation of the terms and descriptions of the components used in machine milking and an acceptable performance of milking machine installations. It has drawn on recent research and devised a standard unifying the many national standards that existed. Furthermore a test procedure has been produced which can be used to assess the acceptability of milking machine systems and to check their performance at regular intervals after installation.

The International Standard, 5707, on Milking Machine Installations was prepared jointly with the International Dairy Federation (I.D.F.) and other interested bodies, ie., European Committee of Associations of Manufacturers of Agricultural Machinery (C.E.M.A.) and the International Committee on Recording the Productivity of Milk Animals (I.C.R.P.M.A.). It specifies the minimum performance requirements and certain dimensional requirements for the satisfactory functioning of milking machines. In addition it lists the requirements of materials, construction and installation. ISO Standard No. 6690 on Mechanical tests was produced at the same time with the object of providing test procedures for confirming that an installation meets the minimum requirements of ISO 5707.

Organisation of Milking Parlour

Milking machines were developed to meet demands for milking more cows more quickly using fewer people and less effort. Initially, this was achieved by the introduction of bucket units which were carried from cow to cow in traditional cowsheds or milking barns. Pipeline milking achieved considerable improvement in labour efficiency and reduction in manual lifting and carrying, but the major development was the change to parlours where the operators use stationary equipment to milk the cows as they pass through the installation during the course of milking.

Installations

At first, static parlour design followed the cowshed stall arrangement with the cows standing side-by-side and a milking unit positioned between each pair of stalls. In these abreast parlours, cows enter across the operator's working area and both are on the same floor level. Later, a step was included to elevate the cows 0.3–0.4 m. Even with this addition milking cannot be carried out in an upright position and it was not until the introduction of the parlour that genuine two-level milking became possible. In these, the cows stand head-to-tail in individual stalls on one or both sides of the operator's pit or work area with a floor level difference of 0.8 m. Each stall is fitted with an entry and exit gate giving access to and from a passage flanking the stalls. A simplified version of the tandem, known as the chute parlour, eliminates the need for separate access passages by having batch entry/exit of cows through the stalls when a division between each stall is opened. The number of cows in each batch equals the number of stalls on each side of the operator's pit. In both the tandem and the chute the distance between udders of adjacent cows is 2.5 m. This disadvantage renders large parlours impracticable, a problem which was solved by the development of the herringbone parlour. By standing a batch of cows in echelon formation at an angle of 30°–35° to the sides of the operator's pit, the distance between udders is reduced to 0.9 m. There are no individual stalls, the cows being restrained on the platform (or standing) by an entry gate, an exit gate and a rump rail parallel to the pit side. Herringbones have become popular in all major milk producing countries, being suitable for herds of 50 to 400 cows. In a recent modification, called the *side-by-side*, the cows stand at right angles to the pit so that 3 cows can occupy the space required for 2 cows in a herringbone.

Four-sided (polygon) and, more recently, three-sided (trigon) herringbones have been built for larger herds. These multi-sided parlours economise in the number of units and stalls required compared with the conventional two-sided herringbone because fewer units are idle at any one time during milking. (eg. a 16/16 trigon is equivalent to a 20/20 herringbone in terms of parlour performance capacity). Also, the smaller batch size for a given number of units means that a slow milking cow has less effect on batch milking time.

Originally, rotary parlours were built for very large herds but more recently smaller ones have been designed to provide an alternative to the herringbone. As in the case of static parlours, the cows stand either side-by-side, ie, rotary abreast; head-to-tail ie, rotary tandem or in echelon formation, ie, rotary herringbone. During milking, cows walk onto a rotating platform singly with the operator standing at the point of entry to prepare the udders for milking and attach the teatcup clusters. The cows leave the platform when rotation brings them opposite the exit passage, the clusters having been removed automatically when milk flow ceased. High capital and maintenance costs, mechanical faults and the introduction of automation into static parlours have all contributed to a declining interest in rotaries.

Types of Milking Parlour

Even though there are several parlour designs and configurations there are only two basic types; those having one milking unit to each pair of stalls (eg. 5 units, 10 stalls), or one unit to each stall (eg. 10 units, 10 stalls). With the exception of rotaries, trigons and polygons, milking parlours can be of either type. In recent years, many 1 unit per 2 stall parlours have been "doubled-up" to the 1 unit per 1 stall version and it is important that the effect of this change is understood, particularly in relation to comparative parlour performance capacity. The operator's work routine time (ie. the time available to carry out the routine jobs on each cow) is unaffected. This is because, in the doubled-up version about 50% of the units will, on average, be idle at any one time, and the content of the work routine will be unchanged. Also unaffected is the available eating time for cows in parlours where batch milking is practised, (eg. the herringbone). Although the average interval between cluster attachment and removal (ie. the available milking time per cow) becomes greater in the doubled up version for a given performance level, this advantage can be exploited only if the operator had previously been waiting for cows to milk out. In terms of performance capacity (ie. available milking time per cow) doubling up a 5/10 herringbone to a 10/10 for example, is equivalent to adding one more unit to the 5/10 to make a 6/12.

Other advantages of one unit per stall installations are that delays caused by slow milking cows can be minimised because the operator can select the sequence of cluster attachment to cater for known differences in the milking out times of cows; the interval between cow preparation and cluster attachment is likely to be more constant and, milk flow can be gravity assisted to pipelines below udder level.

The expense of doubling-up will achieve only a marginally improved throughput created by an increase in the available milking time per cow, improved flexibility in the use of the milking units and a work routine unimpeded by equipment hanging from the centre of the operator's work area.

When a new 1 stall per 1 unit installation is proposed in preference to a 1 unit per 2 stall alternative, these same marginal advantages are relevant together with a small saving in building space which occurs. For example a 10/10 herringbone requires approximately 1 m less length of building than the 6/12 equivalent

Milk Hygiene

The milk secreted into an uninfected cow's udder is sterile. Invariably it becomes contaminated during milking, cooling and storage, and milk is an excellent medium for bacteria, yeasts and moulds that are the common contaminants. Their rapid growth, particularly at high ambient temperatures can cause marked deterioration, spoiling the milk for liquid consumption or manufacture into dairy products. This can be avoided by adopting the simple, basic rules of clean milk production.

Udder Infection

The essential requirements are to maintain udders free from infection (eg. mastitis); manage cows so that their udders and teats are clean; milk them in such a way that minimises bacterial contamination; storc thc milk in clean containers and, wherever possible, at temperatures which discourage bacterial growth until collected. Simple and low-cost husbandry practises enable milk to be produced with a bacterial count of less than 50,000 per ml. The golden rule of clean milk production is that prevention is better than cure.

It is impossible to prevent mastitis infection entirely but by adopting practical routines it can be kept at low levels. Most mastitis is subclinical and although not readily detected by the stockman, it will not normally raise the bacterial count of herd milk above 50,000 per ml. Once the clinical stage is reached, the count may increase to several millions/ml and one infected quarter may result in the milk from the whole herd being unacceptable. It is important to detect clinical cases and exclude their milk from the bulk.

Other Sources of Contamination

Under normal grazing conditions, cows' udders will appear clean and therefore washing and drying will be unnecessary. Otherwise, any visible dirt must be removed using clean, running water, individual paper towels or cloths in clean water to which a disinfectant has been added (eg. sodium hypochlorite at 300 ppm). If udder cloths are used, provide a clean cloth for each cow. After each milking wash and disinfect them and hang up to dry. Disposable paper towels are preferable and more effective for drying after washing. When cows are housed or graze in heavily stocked paddocks, external udder surfaces are usually grossly contaminated with bacteria even when they appear visibly clean, therefore routine udder preparation procedures should be followed. Whenever udders are washed they should be dried.

Foremilking has little affect on the total bacterial count of the milk but is an effective way of detecting clinical symptoms of mastitis. Filtering or straining the milk removes visible dirt but not the bacteria in the milk because they pass through the filter. Aerial contamination of milk by bacteria is insignificant under normal production conditions.

The milk contact surfaces of milking and cooling equipment are a main source of milk contamination and frequently the principal cause of consistently high bacterial counts. Simple, inexpensive cleaning and disinfecting routines can virtually eliminate this source of contamination.

Cleaning Milk Production Equipment

It is virtually impossible with practical cleaning systems to remove all milk residues and deposits from the milk contact surfaces of milking equipment. Except in very cold, dry weather, bacteria will multiply on these surfaces during the interval between milkings, so that high numbers (eg 106 per m2) can be present on visually clean equipment. A proven cleaning and disinfectant routine is required so that with the minimum of effort and expense, the equipment will have low bacterial counts as well as being visually clean.

The essential requirements are, to use milking equipment with smooth milk contact surfaces with minimal joints and crevices, an uncontaminated water supply, detergents to remove deposits and milk residues and a method of disinfection to kill bacteria.

Water Supplies

Unless an approved piped supply is available it must be assumed that water is contaminated and therefore hypochlorite must be added at the rate of 50 ppm to the cleaning water. Hard water (ie. high levels of dissolved calcium and other salts) will cause surface deposits on equipment and reduce cleaning effectiveness. In such cases, it is necessary to use de-scaling acids such as sulphamic or phosphoric, periodically.

Detergents and Disinfectants

Detergents increase the 'wetting' potential over the surfaces to be cleaned, displace milk deposits, dissolve milk protein, emulsify the fat and aid the removal of dirt. Detergent effectiveness is usually increased with increasing water temperature, and by using the correct concentration and time of application. Detergents contain inorganic alkalis (eg. sodium carbonate and silicates and tri-sodium phosphate), surface-active agents (or wetting agents), sequestering (water-softening) agents (eg. polyphosphates) and acids for de-scaling. Many proprietary, purpose-made detergents are usually available, but otherwise, an inexpensive mixture can be made to give a concentration in solution of 0.25% sodium carbonate (washing soda) and 0.05% polyphosphate (Calgon). Disinfectants are required to destroy the bacteria remaining and subsequently multiplying on the cleaned surfaces. The alternatives are either heat applied as hot water or chemicals. Heat penetrates deposits and crevices and kills bacteria, providing that correct temperatures are maintained during the process of disinfection. The effectiveness of chemicals is increased with temperature but even so, they do not have the same penetration potential as heat and they will not effectively disinfect milk contact surfaces which are difficult to clean.

When hot water alone is used, it is best to begin the routine with water at not less than 85°C, so that a temperature of at least 77°C can be maintained for at least 2 minutes. Many chemicals are suitable as disinfectants, some of them combined with detergents (ie. detergent-sterilisers). Use only those which are approved, avoiding particularly those which can taint milk (eg. phenolic disinfectants). *Always follow the manufacturers instructions.* Sodium hypochlorite is an inexpensive example of an approved disinfectant suitable for most dairy purposes. Sodium hydroxide (caustic soda) can also be very effective at concentrations of 3%–5% at ambient temperatures, providing adequate contact time is given with the surfaces to be cleaned and disinfected.

Milking Premises

The milking premises should have a dairy or suitable place equipped with a piped hot and cold water supply, a washtrough, brushes, a work surface, storage racks and cupboards and, if necessary, a vacuum pipeline connection. In addition, it is advisable to have a dairy thermometer (0°C - 100°C), rubber gloves and goggles for use when handling chemicals.

Daily routines for cleaning and disinfecting vary with the size and complexity of the milking installation but will include methods of removing dirt and milk from the equipment followed by disinfection. For hand milking, bucket and direct-to-can milking machines, basic manual methods of cleaning and sterilizing are adequate and effective. For pipeline milking machines in-situ (in-place) systems are necessary.

Milk can become grossly contaminated from bacteria on ancillary equipment which must also be cleaned and disinfected effectively. Coolers, either the corrugated surface or the turbine in-can, can best be cleaned and disinfected manually and stored in the dairy to drain. Refrigerated bulk milk tanks can be cleaned either manually using cold or warm detergent/disinfectant solutions, or for the larger tanks, by automatic, programmed equipment. In either case, a cold water chlorinated (50 ppm) rinse preceeds and follows the washing solution. Foremilk cups can be a potent source of bacterial contamination and need to be cleaned and disinfected after each milking. They should then be stored in the dairy to drain.

It is important with any method of cleaning that the equipment is drained as soon as possible after washing for storage between milkings. Bacteria will not multiply in dry conditions but water lodged in milking equipment will, in suitable temperatures, provide conditions for massive bacterial multiplication. Equipment with poor milk contact surfaces, crevices and large number of joints, remaining wet between milkings in ambient temperatures above 20°C, should receive a disinfectant rinse (50 ppm available chlorine) before milking begins.

Mastitis Control

Mastitis is an inflammation of the udder and is common in dairy herds causing important economic losses. It cannot be eradicated but can be reduced to low levels by good management of dairy cows.

Of the several causes of mastitis only microbial infection is important. Although bacteria, fungi, yeasts and possibly virus can cause udder infection the main agents are bacteria. The most common pathogens are Staphylococcus aureus, Streptococcus agalactiae, Str. dysgalactiae, Str. uberis and Escherichia coli though other pathogens can cause occasional herd outbreaks. Mastitis occurs when the teats of cows are exposed to pathogens which penetrate the teat duct and establish an infection in one or more quarters within the udder. The course of an infection varies, most commonly it persists for weeks or months in a mild form which is not detected by the stockman (ie. subclinical mastitis). With some pathogens, typically E coli, the infection is frequently more acute and there is a general endotoxaemia with raised body temperature, loss of appetite and the cow may die unless supportive therapy is given. When clinical mastitis occurs the effective therapy is a course of antibiotic infusions through the teat duct. These nearly always remedy the clinical disease and often eliminate the bacterial infection. Infections may spontaneously recover but most persist to be eliminated eventually by antibiotic therapy or when the cow is culled. The susceptibility of cows varies considerably and new infections are most common in older cows in early lactation, at the start of the dry period and when the management is poor.

Mastitis causes direct economic losses to farmers in several ways. Milk yields are reduced, milk that is abnormal or contaminated with antibiotics is unsaleable, there are veterinary and antibiotic costs, a higher culling rate and occasional fatalities. The milk processing industry also incurs losses because of problems that result from antibiotic in milk, and the reduced chemical and bacterial quality of mastitic milk.

Infectious Mastitis

The several microbial diseases of the udder that are collectively known as mastitis are distinctly different. The pathogens can arise from different primary sites, they multiply in different environments and therefore the timing of the exposure of the cow to the bacteria will vary. Subsequently the acuteness and persistency of the infections differ and also the probability of cure when therapy is given.

The commonest forms of mastitis in most countries are caused by S. aureus and Str. agalactiae. The primary sites of these is the milk of infected quarters and therefore they are spread mainly at milking, either during udder preparation or on hands and milking machines. These pathogens can colonise and multiply in teat sores and in teat ducts and this greatly increases the degree of exposure of the teats to bacteria. They usually cause chronic infections which persist in the subclinical form and occasionally become clinical when abnormal milk can be detected. Systemic infection with loss of appetite and raised body temperature is infrequent. When suitable antibiotic preparations are infused into the udder the clinical mastitis nearly always subsides and most Str. agalactiae infections are cured but with staphylococcal infections the cure rate is poor and most persist.

Infections caused by Str. uberis and E. coli are often called 'environmental'. The main primary sites of the pathogens are bovine, but not from within the udder. These do not normally colonise teat skin and the multiplication occurs in organic bedding materials (eg. straws and sawdust). These types of infection are most common in housed cattle in early lactation and whilst they can cause persisting subclinical mastitis the more typical from is clinical mastitis soon after the onset of the infection, and with coliform mastitis the endotoxaemia causes raised body temperature and marked reductions in milk production. Str. uberis infections usually respond to therapy but with E. coli infections it is important to give supportive treatment to overcome the endotoxaemia and if this is successful spontaneous recovery usually follows.

Str. dysgalactiae is similar to Str. agalactiae and S. aureus in that it can readily colonise and multiply in teat lesions but the main primary site is not the milk of infected quarters, but other bovine sites. The course of the infection is not dissimilar to Str. agalactiae and infections respond readily to antibiotic therapy.

Many other microorganisms can cause mastitis. These less common forms are not usually important but pseudomonads and Mycoplasma bovis does cause serious problems in a few herds.

Although the pathology of the various types of infection show distinct differences the causes of infection can be diagnosed with certainty only by bacteriological tests made on aseptically taken quarter milk samples.

Methods of Mastitis Control

Mastitis cannot be eradicated but can be reduced to low levels by adopting simple economic management routines that relate to the patterns of infection. Currently it is not possible to base a control on vaccination and although cows susceptibilities are largely inherent there are major practical limitations to control through breeding, and progress would be slow. Therapy is invaluable to contain the problem but cannot be the basis of a control which must depend onpreventing new infections. In practice the key to control is good cattle management particularly steps to reduce exposure to pathogens and also the planned use of antibiotic therapy. Because control depends on management the steps must be simple and economic and fit easily into a milking routine.

The following routine will reduce the proportion of infected cows and clinical mastitis by at least 70% if used regularly at each milking. Mastitis caused by Str. agalactiae will be reduced to very low levels and is frequently eradicated.

1. *Adopt good cow management practices as the essential basis for a mastitis control routine (eg. feeding, housing, hygiene). Mastitis is unlikely to be controlled with neglected, underfed cows kept under stress in dirty conditions.*

2. *Reduce exposure to pathogens*
   - Clean thoroughly all equipment used when milking
   - Do not house or corral cattle under dirty conditions, preferably change organic bedding materials daily or use sand for bedding
   - Wash dirty udders before milking with clean running water preferably with the hand, a disposable paper towel or a disinfected cloth and dry thoroughly. Do not wash with contaminated cloths and water
   - Dip or spray all teats after milking with disinfectant teat dip
   - Adopt practices that prevent the occurrence of teat lesions. If they occur use a teat dip or spray containing a emolient.
   - If practical milk clinically affected cows last
   - Additional benefits can be obtained by disinfecting hands before milking each cow, using individual paper udder cloths, dipping teat cups in disinfectant before each cow is milked, and 'back flushing'.
3. *Reduce the chances of pathogens penetrating the teat duct by*
   - avoiding teat injury or fly attack
   - using a milking machine that is correctly tested, and maintained
   - using a milking machine modified to prevent 'reverse flow' and 'impacts'
   - minimise the effects of vacuum fluctuations
4. *Reduce the duration of infections by*
   - detecting clinical mastitis by examining foremilk or fitting 'mastitis detectors' into the long milk tubes
   - giving intramammary infusions of antibiotics under veterinary supervision to clinically affected cows
   - Treating cows at drying off with infusions of antibiotics recommended by a veterinarian
5. *Reduce mastitis in nonlactating growing cattle or cows in the dry period*
   - Avoid using low lying grazing land and damp wooded areas where flies are common. Move cattle from pastures known to give problems with mastitis

- Adopt good fly control measures
- Treat cows at drying off with antibiotics recommended by veterinarian. All cows should be treated, alternatively treat cows that have previously shown signs of infection

Note the reduction in infection is not immediate but levels fall by about 50% in one year and continue to fall in successive years.

References

Hogeveen, H.,W., et al., (2001), "Milking interval, milk production and milk flow-rate in an automatic milking system", *Livestock Production Science*, Vol. 72, pp. 157–167.

Hopster, H., et al., (2002), "Stress Responses during Milking; Comparing Conventional and Automatic Milking in Primiparous Dairy Cows", *Journal of Dairy Science* Vol. 85, pp. 3206–3216.

Millar, K. M., (2000), "Respect for Animal Autonomy in Bioethical Analysis: The Case of Automatic Milking Systems (AMS)",*Journal of Agricultural and Environmental Ethics*, Springer, Netherlands, Vol. 12, No. 1, pp. 41–50.

Rossing, W. and Hogewerf, P. H., (1997), "State of the art of automatic milking systems", *Computers and electronics in agriculture*, Vol. 17, pp. 1–17.

Schukken, Y.H., Hogeveen H., and Smink, B.J., (1999), "Robotic Milking and Milk Quality, Experiences From the Netherlands",*National Mastitis Council Regional Meeting Proceedings 1999*, pp. 64–69.

9

Animal Slaughtering and Meat Processing

The activities of the meat sector may be divided into three stages - slaughtering, meat cutting and further processing. Each stage involves completely different technical operations which must not be viewed as separate and independent processes. There are significant interactions between the stages and shortcomings at one stage can have a serious negative impact on the product or process in a subsequent stage. They may influence technological, biochemical or microbiological aspects.

Improper slaughtering techniques such as faulty stunning, bleeding, skinning, evisceration and carcass splitting can damage parts of the carcass and certain byproducts and make them unsuitable for further use. Poor standards of hygiene during slaughtering or carcass handling result in high levels of mirobial contamination in the meat, thus reducing the shelf-life and adversely affecting the sensoric properties of products fabricated from this raw material. Although controls imposed on the meat industries have become more stringent and effective, improper treatment of slaughter animals and poor meat-handling techniques persist in many meat plants. These problems are evident in many developing countries. Apart from deficiencies in veterinary meat inspection, which is not the subject of this publication, serious shortcomings with regard to general meat hygiene and meat technology can frequently be observed. This is to some extent due to the lack of adequate facilities in the meat sector in developing countries, but carelessness and lack of skills on the part of the personnel involved in meat operations are also important factors.

General Hygiene Rules

It is essential that all meat-processing operations, whether slaughtering, cutting or further processing, be carried out in a clean area and, as much as possible, that the products be protected from contamination from all sources.

When meat-processing operations are carried out within a facility specifically built and maintained for meat processing, sources of contamination can be much more easily and adequately controlled. The following requirements are considered essential to good sanitary preparation of meat and meat products.

Facilities

Floors. Brick, tile, smooth concrete or other impervious, waterproof materials are suitable for floors. In some areas wooden floors will suffice if they are tight, smooth, in good repair and properly maintained. Wooden floors are not suitable in areas where slaughtering or curing takes place and meat juices and moisture collect.

Drains. To carry away waste liquids, there should be sufficient drains of the proper size that are correctly located, trapped and vented. All floors should be sloped toward the drains. Generally for adequate waste disposal, one drain is needed for each 18 m_2 of floor space in slaughtering areas, and one drain for each 46m_2 in processing and other areas.

Walls. Glazed tile, smooth cement plaster, rustproof metal panels and smooth plastic panels that are properly caulked are all acceptable for walls in processing and refrigerated areas because they can all be effectively cleaned and sanitized. Other materials are also acceptable if they can be satisfactorily cleaned. In no instance should walls be made of materials that absorb moisture or other liquids. Ceilings must be tight, smooth and free from any scaling that may fall into the meat products, and should also be of moisture-resistant materials. All light bulbs should be covered with unbreakable material to prevent broken pieces from falling into the product.

Doors and doorways. All doorways through which the product must pass, whether suspended on rails or lying on hand trucks, should be wide enough to ensure that the meats never touch the doorways risking contamination. Wooden doors and doorways should be covered with metal with tightly soldered seams.

Water supply. Whether from individually owned and controlled sources such as wells or streams or from a municipal system, the water supply must be potable and abundant cold and hot water must be distributed to all parts of the operation.

Lighting. In all areas where products are critically examined during sanitary control or for cleanliness, 50-foot candles of light should be provided. For adequate visibility 20-foot candles of light should be provided wherever any processing occurs. In all other areas, such as dry storage, there should be sufficient light to keep the area orderly and sanitary.

Refrigeration. The main purpose of refrigeration is to cool the meat down after slaughter and to maintain it in a chilled state for shorter or longer storage periods and for cutting and further processing. If frozen storage is provided and utilized, it should be maintained at the lowest possible temperature for maximum shelf-life. Minus 18° to -12°C is satisfactory freezer storage; however, large quantities of product must either be quickfrozen prior to storage or thinly spread out to facilitate freezing. It is also recommended that all rooms where meat is processed, except in the slaughter and cooler storage areas, should be maintained at a temperature of about 12°C. In facilities where no refrigeration or cooling is furnished in processing areas, the handling of meat products is possible if all equipment contacting the products is throughly cleaned and sanitized from time to time (recommended every four hours). Frequent cleaning is necessary because in warmer temperatures bacteria multiply rapidly and the risk of product contamination increases.

Equipment

The equipment needed for converting livestock into meat products need not be elaborate and expensive. The amount of equipment will depend on the slaughtering and processing procedures employed. If possible, all equipment should be made of stainless steel or plastic, be rust resistant and easily cleaned and sanitized.

All equipment should be constructed of stainless steel, galvanized steel, aluminium or approved plastic. Wooden tables are not acceptable because wood absorbs meat juices and fats and cannot be thoroughly cleaned. Hardwood cutting-boards maintained smooth and free from checks and cracks may be used. Cutting tables covered with other than hard plastic are not acceptable for contact with meat.

All other equipment should be of the type that can be taken apart and thoroughly cleaned. Any stationary equipment must be located far enough from walls to permit proper cleaning around and under it.

In all areas there should be conveniently located foot-pedal or kneeoperated wash-basins with hot and cold water, soap and disposable towels. In slaughtering areas, lavatories should be convenient to the dressing operations. Hot-water containers, either electric or steam-heated to 82°C, should be available for sanitizing tools contaminated with diseased material or other filth during dressing.

Rails must be located high enough to prevent meat from touching the floor. For beef carcasses, the minimum height for rails should be 3.4 metres, while 2.4 metres is sufficiently high for small livestock such as goats, hogs and sheep. Rails should also be far enough away from fixed objects and walls to avoid contact.

Personnel Hygiene

Probably as important as anything in the production of clean, wholesome, unspoiled products is the attitude of the workers toward cleanliness. Personnel with clean hands, clothing and good hygienic practices are absolutely essential to the production of high-quality foods.

All clothing should be clean, in good repair and made of washable material. Street clothing should be covered with coats or gowns while handling exposed product. White or light-coloured clothing is most desirable and garments that become soiled or contaminated should be changed when necessary.

All persons working with exposed meat products should have their hair under control, either completely covered with a clean cap or hat or confined by a hairnet to prevent hair from falling into products.

Safety devices such as aprons, wrist guards and mesh gloves must be made of impervious material, clean and in good repair. At no time should leather aprons, wrist guards or other devices be worn unless clean, washable coverings are used over them. Light-coloured rubber or plastic gloves may be worn by product handlers only if clean and in good repair.

No person working with meats should wear any kind of jewellery, badges or buttons that may come loose and be accidently included in the product.

Shoes and boots should be worn at all times and should be appropriate for the operations being conducted. They should also be made of impervious materials. Any aprons, knives and footwear that become contaminated during operations should be routinely cleaned in areas or facilities provided for that purpose.

No cloth twine, belts or other similar materials should be used to cover implement handles or used in other places where they may harbour filth and serve as a ready source of product contamination.

All unsanitary practices should be avoided by meat handlers. No one should smoke or use tobacco in areas where edible products and ingredients are handled, prepared or stored, or where equipment and utensils are cleaned. When handling edible products, scratching the head, placing fingers in or around the nose or mouth, sneezing or coughing on the product should never occur. Workers must also guard against contaminating products from localized infections or sores.

Workers can contaminate carcasses and meat through handling, coughing and sneezing. This may cause rapid spoilage of the meat or, more seriously, food poisoning. Coughs and sneezes are a particularly effective way of transmitting bacteria to meat. Transfer of faecal matter either of animal or human origin to the meat is particularly hazardous. Most contamination on the hands of workers in slaughter floors with faecal matter comes from the hides and fleeces.

Hands should be washed frequently to remove all visible soiling. Stainless-steel sinks without plugs should be conveniently accessible to all workers. Water should be supplied at approximately 43°C to a simple tap which is foot- or knee-operated. Liquid disinfectant soap and paper towels should be available. Particular attention should be paid to cleaning under the fingernails. Hands should also be thoroughly washed after using the toilet, smoking, coughing or sneezing, handling money, garbage or soiled or infected material.

Routine Cleaning of Rooms and Equipment

The floors should be kept clear of all debris, such as hooves and horns, in slaughterhalls or other inedible parts or fat and meat particles in cutting, processing and by-product handling areas, and must be frequently washed down. At the end of each day a thorough cleaning programme should be followed. All matter should be removed from floors, platforms, gullies, etc., followed by a thorough hosing down of walls, floors and all surfaces to loosen dirt. Finally a strong cleaning solution should be applied and left for a while before being rinsed off. A thorough inspection should be made afterwards and any areas remaining soiled should be cleaned again.

In order to maintain the cleanest possible products a standard cleaning routine of the equipment should be established. Initially all large pieces of refuse material should be scraped or swept together and disposed of. Follow-up should include scrubbing of the equipment using brushes and a soap or detergent and a complete sanitizing with hot water at 82°C and an approved chlorine or iodine rinse. Finally, a coating of light mineral oil can be applied to metal equipment, particularly that not fabricated of stainless steel, to prevent rust.

Principles For Meat Handling

Current recommendations for handling all meat products are to keep them clean, cold and covered in order to maintain quality and protect against food poisoning and disease. Generally contamination occurs when the product comes into contact with dirty hands, clothing, equipment or facilities. If the product is kept clean there will be little or no contamination by microorganism.

Effect of bacterial growth on the shelf-life of meat and meat products

The total viable count of bacteria (TVC) expressed as organisms/cm^2 or as organisms/g on fresh meat or a meat product sets a limit to its shelf-life. Meat will "spoil" with TVC at 10^6/cm^2 because of off-odours. Slime and discoloration appear at 10^8/cm^2. The main factors determining the time taken for the TVC to reach these levels are the initial count due to contamination during slaughtering and processing, further contamination during storage, temperature, pH and relative humidity. An example of how the level of contamination affects shelf-life is shown in Table 1.

Table 1: Effect of initial contamination on the storage life of lean beef

Initial bacterial count (org./cm^2)	*Days at 0°C before slime development*
100 000	8
10 000	10
1 000	13
100	15
10	18

After cleanliness, keeping meat products cold is the second most important requirement in order to achieve a desirable shelf-life. Microorganisms rapidly proliferate at elevated temperatures and slime development is a definite visual sign of microbial growth. The importance of temperature in the control of microbial growth is shown in Table 2.

Table 2: Relationship between storage temperature and slime development

Storage temperature (°C)	***Days before slime develops***
0	10
1	7
3	4
5	3
10	2
16	1

Bacteria relevant to meat, meat products and other food are divided into three groups according to the temperature range within which they can grow: mesophiles 10–45°C, psychrophiles 0–28°C and psychrotrophs 10– 45°C, or slow growth at 0–10°C. Mesophiles will not grow below 10°C but psychrotrophs, of which Pseudomonas are the more important, will grow down to 0°C. The nearer to 0°C the storage temperature the slower the growth of the spoilage bacteria and the longer the shelf-life.

Under ideal conditions bacteria double in number every 20 minutes. A single bacterium multiplies to over one million in less than seven hours.

Some bacteria cause product spoilage, others cause food poisoning. The former limit product shelf-life but the latter cause illness. Almost all foodpoisoning bacteria are mesophiles so refrigeration below 10°C offers good protection. Many mesophiles cause spoilage, but since meat is refrigerated most spoilage is due to psychrophiles. Storing meat at temperatures close to 0°C will inhibit the growth of pyschrotrophs. Shelf-life will be extended by avoiding contamination through good hygiene practices.

Effect of Contamination on Sensoric Properties of Meat

Aerobic spoilage by bacteria and yeasts usually results in slime formation, undesirable odours and flavours (taints). Colour changes, rancidity, tallowy or chalky flavours from the breakdown of lipids may also occur. Colour changes as a result of pigment oxidation may be grey, brown or green discoloration. Aerobic spoilage by moulds results in a sticky surface, musty odours, alcohol flavours and creamy, black or green discoloration. Anaerobic spoilage which occurs either within the meat or on the surface in sealed containers where oxygen is absent or very limited is marked by a souring due to the production of organic acids and gases.

Food Poisoning

Food poisoning may be due to infection or intoxication. Infection is caused by the consumption of live bacteria which multiply in the body producing characteristic symptoms. Intoxication is due to toxins in food produced by bacteria before the food was eaten. Toxins are chemical compounds which may linger in food with no microbes growing in it, and are therefore very dangerous.

Salmonellae are facultative anaerobes which cause infectious food poisoning. Ten or 20 cells of *Salmonella typhi* are sufficient to cause typhoid but 10 000 to 100 000 cells of other species may be necessary to cause an infection. Some are host-specific affecting the animal from which the meat was produced but failing to cause infection when consumed by man. Typical symptoms of salmonellosis include diarrhoea, fever and vomiting. The illness may last one to 14 days after a 12 to 24-hour incubation period. Victims may excrete the bacteria for weeks after the symptoms subside. Poor personal hygiene will cause contamination of meat.

Staphylococcus aureus is a facultative aerobe that causes intoxication. It lives in the nose, throat, hair and skin and on animal hides. Meat is contaminated by handling and by sneezing or coughing. Minute amounts of the toxin will cause illness, which starts within one to eight hours of eating poisoned food. Nausea, vomiting and shock may last for one to two days. On rare occasions it is fatal. This bacterium does not produce off-odours or spoilage so it cannot be easily checked. Refrigeration will control its growth. Cooking may destroy the bacteria but not the toxin as it is heat stable. It is particularly troublesome in cooked cured meats, normally as a result of recontamination after the curing process in subsequent handling, for instance during slicing.

Clostridium botulinum, an anaerobe, produces the toxin botulin, one of the most poisonous substances known. This attacks the central nervous system causing death by respiratory paralysis. Dormant cells occur everywhere in the soil, fish, animals and plants. High-moisture, low-acid, low-salt conditions at above 3°C favour growth and toxin production. Control measures must destroy spores or prevent growth and toxin formation. Botulism is usually due to undercooking processed meats. Pressure-cooking will give commercial sterility. Pasteurization (heating to 70°C) and adding salt (NaCl) and sodium nitrite ($NaNO_2$) is used for canned ham. Refrigeration (0–10°C) is essential for vaccum-packed meats. Frozen storage prevents growth.

Clostridium perfringens, an anaerobic bacterium, is a common cause of food poisoning but is rarely fatal. It grows well in warm meats so is usually found in left-over meats that have not been kept chilled and not been reheated to 70°C to kill the bacteria present. The main symptoms are diarrhoea and weakness which last for 12 to 24 hours after an incubation period of eight to 20 hours.

Hygiene Practices in Slaughtering and Meat Handling

Equipment

Slaughtering equipment, particularly for smaller-scale operations, need not be elaborate and expensive. The amount of equipment will depend on the slaughtering procedures employed. If possible, all equipment should be made of stainless steel or plastic, be rust resistant and easily cleaned and sanitized. Equipment which does not get in contact with the meat (e.g. overhead rails, working platforms, knocking pen) is usually made of galvanized steel.

Basic equipment needed for the slaughtering operation:

- stunning gun, electrical head tongs or simple stunning equipment for direct blow
- knives:

  sticking - 15 cm sharpened on both sides

  skinning - 15 cm curved
- a sharpening steel
- oil or water sharpening stone
- scabbard and belt for holding knives
- meat saw - hand or electric and cleaver
- block and tackle or chain hoist strong enough to hold the weight of the animal to be slaughtered

- pritch, chocks or skinning rack (dressing cradle)
- a strong beam, tripod or track 2.4 to 3.4 m from floor
- spreader - gambrel or metal pipe
- several buckets
- working platforms
- scalding barrel or tank
- pot, barrel or system for boiling water
- bell scrapers
- solid scraping table or platform
- thermometer registering up to 70°C
- hog or hay hook
- torch or flame for singeing

The last seven items indicate additional equipment required when hogs are scalded and scraped rather than skinned.

Useful additional equipment:

- knocking pen
- bleeding hooks (for vertical bleeding)
- blood-catching trough
- wash trough (tripe) Sanitation of hands and tools:
- hand wash-basin
- implement sterilizers

Means should be available to clean thoroughly all equipment coming into contact with carcasses or meat. Implement sterilizers are stainless-steel boxes holding hot (82°C) water, shaped to suit particular equipmentknives, cleavers, saws, etc. Knife sterilizers should be placed in positions where every operator who uses a knife has immediate access. Handles as well as blades must be sterilized. Each operator should have at least two knives etc., one to use while the other sterilizes.

Treatment of Livestock before Slaughter

Stress in its many forms, e.g. deprivation of water or food, rough handling, exhaustion due to transporting over long distances, mixing of animals reared separately resulting in fighting, is unacceptable from an animal welfare viewpoint and should also be avoided because of its deleterious effects on meat quality. The most serious consequence of stress is death which is not uncommon among pigs transported in poorly ventilated, overcrowded trucks in hot weather. From loading on the farm to the stunning pen animals must be treated kindly, and the lorries, lairages and equipment for livestock handling must be designed to facilitate humane treatment. Stress immediately prior to slaughter, such as fighting or rough handling in the lairage, causes stored glycogen (sugar) to be released into the bloodstream. After slaughter this is broken down in the muscles producing lactic acid. This high level of acidity causes a partial breakdown of the muscle structure causing the meat to be pale, soft and exudative (PSE). This condition is mostly found in pigs.

Long-term stress before slaughter such as a prolonged period of fighting during transport and/or lairage leads to exhaustion. The sugars are used up so that less is available to be broken down and less lactic acid is produced.

The reduced acidity leads to an abnormal muscle condition known as dark, firm and dry (DFD) in pigs or dark cutting in beef. The condition is rarer in lamb. Such meat has a high pH (above 6.0) and spoils very quickly as the low acidity favours rapid bacterial growth.

Handling Animals during Transport and Lairage

An electric goad should be used rather than a stick or tail-twisting not only to avoid stress but also to prevent carcass bruising. Grabbing sheep by the fleece also causes bruising.

To avoid fighting, animals not reared together must not be mixed during transport and lairage. Load and unload using shallow stepped ramps to avoid stumbles. Trucks should be neither over- nor underloaded. Overloading causes stress and bruising due to crushing. Underloading results in animals being thrown around and falling more than necessary. Drivers should not corner at excessive speed and must accelerate and decelerate gently.

The lairage should have small pens. Corridors must curve and not bend sharply so that stock can see a way forward. Stock must not be slaughtered in sight of other stock. Plenty of clean water must be available. The lairage must be well lit and ventilated. Do not hold stock in lairage for more than a day. Only fit, healthy stock may be slaughtered for human consumption.

Fasting before slaughter reduces the volume of gut contents and hence bacteria and therefore reduces the risk of contamination of the carcass during dressing. It is usually sufficient for the animals to receive their last feed on the day before slaughter. Stock should have a rest period after arrival at the slaughterhouse. However, long periods in the lairage can lead to DFD if the animals are restless and fighting or mounting.

Animals should be as clean as possible at slaughter. Producers should wash their animals before leaving the farm. Trucks used for transport must be washed after each load and the lairage at the slaughterhouse should be kept clear of faecal matter and frequently washedh.

Stunning and Bleeding of Slaughter Animals

Stunning Prior to Bleeding

Most countries have legislation requiring that animals are rendered unconscious (stunned) by a humane method prior to bleeding. Exceptions are made for religions which require that ritual slaughter without prior stunning is practised, provided the slaughter method is humane. Stunning also makes sticking (throat-slitting) less hazardous for the operator. The animal must be unconscious long enough for sticking to be carried out, and for brain death to result from the lack of blood supply.

Methods of Stunning

Direct blow to skull using a club or poleaxe. The blow must be dealt with precision and force, so that the skull is immediately smashed, causing instantaneous unconsciousness. In cattle the aiming point is in the middle of the forehead in line with the ears, where the skull is thinnest. Horses have thinner skulls and are therefore easier to stun by this method. In sheep and goats the brain is more easily reached from the back of the neck. Pigs have a well-developed frontal cavity so the blow should be aimed slightly above the eyes.

Slaughtering mask. A bolt held in the correct position by the mask is driven into the animal's brain by a hammer blow. The device is usually fitted with a spring which returns the bolt to its original position.

Free bullet fired from a pistol into the skull is effective but unsafe. This method has been used on horses and cattle.

Captive-bolt pistols fitted with a blank cartridge are effective on cattle and sheep but not pigs whose skulls are thicker. After firing, the bolt returns to its original position in the pistol. The bolt may or may not be designed to penetrate the skull. With penetrating types the brain becomes contaminated with hair, dirt and bone fragments. If brains are to be saved as edible tissue then the non-penetrating type with a mushroom-shaped head should be used.

Electrical stunning. An electric current of high frequency but, in the case of manually operated equipment, of relatively low voltage (60–80 V) is passed through the brain of an animal for a few seconds to produce unconsciousness. If applied correctly a deep state of unconsciusness is invariably achieved. Strict safety rules must be observed. Head tongs are suitable for pigs and sheep but not for cattle. The electrodes carried on the ends of the tongs must be accurately placed. Places where the skull is thick must be avoided. Electrical contact is impeded by hair and caked mud. Water or brine will improve contact but the head must not be completely wet otherwise the current will have a short-circuit path avoiding the brain. The electrodes must be applied with strong pressure.

Carbon dioxide stunning is used only in large pig abattoirs. Pigs are induced into a chamber and exposed to a concentration of 85 percent CO_2 for about 45 seconds. Although effective for anaesthetizing sheep, it is impractical because of large amounts of CO_2 collecting in the wool and affecting operators on the killing line.

Bleeding after Stunning

The objectives of bleeding are to kill the animal with minimal damage to the carcass and to remove quickly as much blood as possible as blood is an ideal medium for the growth of bacteria.

Sticking, severing the major arteries of the neck, should immediately follow stunning. Care must be taken not to puncture the chest cavity or it will fill with blood.

Cattle. Insert the sticking knife carefully just above the breastbone at 45° pointed toward the head. Ensure that the carotid arteries and jugular veins are severed in one movement.

Sheep. Draw the knife across the jugular furrow close to the head severing both carotid arteries. Alternatively, the knife may be inserted through the side of the neck, though this requires more skill.

Pigs. As for cattle but do not go in too far or a pocket of blood will collect at the shoulder. To reduce contamination by the scalding tank water the cut should be as small as possible.

Bleeding on a Rail

The most hygienic system of bleeding and dressing is to shackle the animal immediately after stunning, then hoist it on to a moving rail. The animal is stuck while being hoisted to minimize the delay after stunning. Bleeding continues until the blood flow is negligible when carcass dressing should begin without further delay,

Blood for human use must be collected with special equipment to avoid contamination from the wound, the gullet of the knife. A hollow knife directs blood away from the wound into a covered stainless-steel container without touching the skin or hide. The knife may be connected to a hose to reduce the risk of contamination. The hose may even be connected to a pump to speed the blood flow. Between 40 and 60 percent of the total blood volume will be removed though this will be reduced if sticking is delayed. To prevent coagulation, citric acid solution made up with one part citric acid to two parts water is added at a rate up to 0.2 percent of the blood volume. The main sources of contamination during sticking and bleeding include the knife, the wound and the food-pipe. The knief should be changed after each operation and returned to a sterilizer. Cutting the hide of sheep and cattle and opening out to make a clean entry for the sticking knife reduces contamination from the wound. If the food-pipe is pierced semi-digested food may be regurgitated contaminating the blood and neck wound.

Horizontal Bleeding

Horizontal bleeding is claimed to give faster bleeding rates and a greater recovery of blood. This may be due to certain organs and blood vessels being put under pressure when animals are hoisted, thus trapping blood and restricting the flow. Bleeding on the floor is very unhygienic. The operation should take place on a specially designed, easily cleaned stainless-steel table which should be cleaned frequently. If blood is to be saved it must not come in contact with the table before reaching the collecting vessel.

Bleeding without Stunning

The Jewish and Muslim religions forbid the consumption of meat which was killed by any method other than bleeding. Since it is difficult to guarantee that all animals will recover consciousness after being stunned by any particular method, stunning is not generally allowed. There are exceptions, however. Some communities do accept low-voltage electrical stunning.

Because animals are fully conscious at the time of sticking, ritual slaughter may be less humane than sticking after stunning. To reduce the suffering operators must be highly skilled so that a successful gash cut severing all the veins and arteries is made quickly at the first attempt. Different communities have different regulations as to the orientation of the animal at sticking, some favouring a position lying on its side, others insisting it lie on its back. The animal should not be hoisted until unconsciousness due to lack of blood supply to the brain is complete.

Scalding and Dehairing of Pigs

Scalding in water at around 60°C for about six minutes loosens the hair in the follicle. Too low a temperature and the hair will not be loosened and too high a temperature and the skin will be cooked and the hair difficult to remove. The simplest equipment consists of a tank into which the pig is lowered by a hoist. The water is heated by oil, gas, electricity or an open steam-pipe.

To check the effectiveness of the scald, rub the skin with the thumb to see if hair comes away easily. Some machines have the thermostatic controls and timers. To reduce contamination, scalding water should be changed frequently, pigs should be as clean as possible at sticking, and bleeding should be fully completed before immersion.

In large factories pigs are transported through scalding tanks with rotating bars or through long scalding tanks stretching from the sticking point to the dehairing point in the time required for an effective scald.

Dehairing is done with a specially formed scraper (bell scraper or knife). If the scald is effective all the hair can be removed by this manual method. Another simple method is to dip the pig in a bath containing a hot resin adhesive. The pig is removed from the bath and the resin allowed to set partially when it is peeled off pulling the hair with it from the root. This is less labour-intensive than scraping and produces a very clean skin. After use the adhesive is melted again, strained to remove the hair and returned to the tank.

Another method of removing dirt and hair in one operation is to skin the carcass though this is only done when the skin is required for leather goods.

With the simple scalding tank, dehairing and scalding may be combined in one operation. Inside the tank are rotating rubber-tipped paddles which are started after closing the lid. As the hair is loosened by the scalding water it is removed by the rubbing effect of the paddles against the skin.

Singeing removes any remaining hairs, shrinks and sets the skin, decreases the number of adhering micro-organisms and leaves an attractive clean appearance. It may be done with a hand-held gas torch. Automated systems transport the pig into a furnace and leave it long enough for an effective singe.After singeing, black deposits and singed hairs are scraped off and the carcass is thoroughly cleaned before evisceration begins.

Skinning of Cattle and Small Ruminants

Cattle

The outer side of the hide must never touch the skinned surface of the carcass. Operators must not touch the skinned surface with the hand that was in contact with the skin.

Combined Horizontal/Vertical Methods

Head. After bleeding, while the animal is still hanging from the shackling chain, the horns are removed and the head is skinned. The head is detached by cutting through the neck muscles and the occipital joint. Hang the head on a hook. Lower the carcass on its back into the dressing cradle.

Legs. Skin and remove the legs at the carpal (foreleg) and tarsal (hind leg) joints. The forelegs should not be skinned or removed before the carcass is lowered on to the dressing cradle or the cut surfaces will be contaminated. The hooves may be left attached to the hide.

Flaying. Cut the skin along the middle line from the sticking wound to the tail. Using long firm strokes and keeping the knife up to prevent knife cuts on the carcass, skin the brisket and flanks, working backwards toward the round. Skin udders without puncturing the glandular tissue and remove, leaving the supermammary glands intact and attached to the carcass. At this point raise the carcass to the half-hoist position, the shoulders resting on the cradle and the rump at a good working height.

Vertical Methods

High-throughput plants have overhead rails which convey the carcass from the sticking point to the chills. Hide removal is carried out on the hanging carcass. The operations are as in the combined horizontal/vertical method, but as it is not possible to reach the hide from ground level more than one operator is needed. A single operator may work with a hydraulic platform which is raised and lowered as required.

Automatic hide pullers are used in high-throughput slaughterhouses. Some types pull the hide down from the hind, others from the shoulders upwards toward the rump.

Automation of hide removal reduces contamination since there is less handling of the carcass and less use of knives. Moving overhead rails also improve hygiene by reducing carcass contact with operators, equipment such as dressing cradles and with each other since carcasses are evenly spaced.

Small Ruminants

Sheep fleeces can carry large volumes of dirt and faeces into the slaughterhouse. It is impossible to avoid contamination of sheep and lamb carcasses when the fleece is heavily soiled. The fleece or hair must never touch the skinned surface, neither must the operator touch the skinned surface with the hand that was in contact with the fleece.

Combined Horizontal/Vertical Method

The animal is turned on its back and cuts are made from the knuckles down the forelegs. The neck, cheeks and shoulders are skinned. The throat is opened up and the gullet (food-pipe) is tied off. The skin on the hind legs is cut from the knuckles down to the tail root. The legs are skinned and the sheep is hoisted by a gambrel inserted into the Achilles tendons. A rip is made down the midline and skinning proceeds over the flanks using special knives or the fists. The pelt is then pulled down over the backbone to the head. If the head is for human consumption it must be skinned or it will be contaminated with blood, dirt and hairs.

Moving cratch and rail system. The hanging carcass is lowered on to a horizontal conveyor made up of a series of horizontal steel plates, bowed slightly and divided into sets large enough to cradle a single animal. Two operators usually work together on each lamb performing the legging operations and opening the skin to the stage where it can be pulled off the back. When the gambrel is inserted into the hind legs it is hoisted on to a dressing rail.

Vertical Method

At sticking the animal is shackled by one hind-leg and left to bleed. Dressing commences with the free leg which is skinned and the foot removed. A gambrel is inserted into this leg and hung on a runner on a dressing rail. The second leg is freed from the shackle, skinned and dressed, then hooked on to the other end of the gambrel. The skin is opened down the midline and cleared from the rump.

A spreader frame (a bar U-shaped at each end) spreads the front legs to simplify work on the neck, breast and flanks. The front toes are held in each end of the frame which is then slung up on to a separate travelling hook. The animal is therefore suspended by all four legs belly uppermost. Skinning continues as in the combined horizontal/vertical method. To clear the shoulders and flanks, the forelegs are freed from the spreader and the feet removed, the animal returning to a vertical position. The skin can now be completely pulled off, including the head if this is for consumption, though this takes some work with the knife. In both methods, after fleece removal the vent and food-pipe are cleaned and tied off.

Evisceration

With all species care must be taken in all operations not to puncture the viscera. All viscera must be identified with the carcass until the veterinary inspection has been passed. After inspection the viscera should be chilled on racks etc. for better air circulation.

Cattle

The brisket is sawn down the middle. In the combined horizontal/ vertical system this is done with the animal resting on the cradle. The carcass is then raised to the half-hoist position and when hide removal is complete the abdominal cavity is cut carefully along the middle line. The carcass is then fully hoisted to hang clear of the floor so that the viscera fall out under their own weight. They are separated into thoracic viscera, paunch and intestines for inspection and cleaning. If any of the stomachs or intestines are to be saved for human consumption, ties are made at the oesophagus/ stomach, stomach/duodenum boundaries, the oesophagus and rectum having been tied off during hide removal. This prevents cross-contamination between the paunch and the intestines.

Small Ruminants

A small cut is made in the abdominal cavity wall just above the brisket, and the fingers of the other hand are inserted to lift the body wall away from the viscera as the cut is continued to within about 5 cm of the cod fat or udder.

The omentum is withdrawn, the rectum (tied off) loosened, and the viscera freed and taken out. The food-pipe (tied off) is pulled up through the diaphragm. The breastbone is split down the middle taking care not to puncture the thoracic organs which are then removed.

Pigs

Loosen and tie off the rectum. Cut along the middle line through the skin and body wall from the crotch to the neck. Cut through the pelvis and remove the bladder and sexual organs. In males the foreskin must not be punctured as the contents are a serious source of contamination. All these organs are considered inedible.

Remove the abdominal and thoracic viscera intact. Avoid contact with the floor or standing platform.

The kidneys are usually removed after the carcass has been split down the backbone. The head is usually left on until after chilling.

Splitting, Washing and Dressing of Carcasses

Hygienic Carcass Splitting with Simple Equipment

Cattle

Work facing the back of the carcass. Split the carcass down the backbone (chine) with a saw or cleaver from the pelvis to the neck. Sawing gives a better result but bone dust must be removed. If a cleaver is used, it may be necessary to saw through the rump and loin in older animals. The saw and cleaver should be sterilized in hot (82°C) water between carcasses. Power saws increase productivity.

Pigs

These are suspended and are split down the backbone as for cattle, but the head is generally left intact.

Sheep

Sheep and lamb carcasses are generally sold entire. If necessary they can be split by saw or cleaver, but a saw will probably be necessary for older animals.

Carcass Washing

The primary object of carcass washing is to remove visible soiling and blood stains and to improve appearance after chilling. Washing is no substitute for good hygienic practices during slaughter and dressing since it is likely to spread bacteria rather than reduce total numbers. Stains of gut contents must be cut off. Wiping cloths must not be used.

Carcass spraying will remove visible dirt and blood stains. Water must be clean. Soiled carcasses should be sprayed immediately after dressing before the soiling material dries, thus minimizing the time for bacterial growth. Under factory conditions bacteria will double in number every 20 or 30 minutes.

In addition to removing stains from the skinned surface, particular attention should be paid to the internal surface, the sticking wound and the pelvic region.

A wet surface favours bacterial growth so only the minimum amount of water should be used and chilling should start immediately. If the cooler is well designed and operating efficiently the carcass surface will quickly dry out, inhibiting bacterial growth.

Bubbling of the subcutaneous fat is caused by spraying with water at excessively high pressure, which may be due to the pressure in the system or a result of holding the spray nozzle too close to the carcass.

Carcass Dressing

The object of carcass dressing is to remove all damaged or contaminated parts and to standardize the presentation of carcasses prior to weighing. Specifications will differ in detail for different authorities. Veterinary inspection of carcasses and offal can only be carried out by qualified personnel. Where signs of disease or damage are found the entire carcass and offal may be condemned and must not enter the food chain, but more often the veterinarian will require that certain parts, for instance those where abscesses are present, be removed and destroyed. Factory personnel must not remove any diseased parts until they have been seen by the inspector otherwise they may mask a general condition which should result in the whole carcass being condemned. Any instructions from the inspector to remove and destroy certain parts must be obeyed.

Refrigeration of Carcasses

Carcasses should go into the cooler as soon as possible and should be as dry as possible. The object of refrigeration is to retard bacterial growth and extend the shelf-life. Chilling meat post-mortem from 40°C down to 0°C and keeping it cold will give a shelf-life of up to three weeks, provided high standards of hygiene were observed during slaughter and dressing.

Carcasses must be placed in the cooler immediately after weighing. They must hang on rails and never touch the floor. After several hours the outside of a carcass will feel cool to the touch, but the important temperature is that deep inside the carcass. This must be measured with a probe thermometer (not glass), and used as a guide to the efficiency of the cooling.

The rate of cooling at the deepest point will vary according to many factors including the efficiency of the cooler, the load, carcass size and fatness. As a general guide a deep muscle temperature of 6–7° C should be achieved in 28 to 36 hours for beef, 12 to 16 hours for pigs and 24 to 30 hours for sheep carcasses. Failure to bring down the internal temperature quickly will result in rapid multiplication of bacteria deep in the meat resulting in off-odours and bone-taint.

High air speeds are needed for rapid cooling but these will lead to increased weight losses due to evaporation unless the relative humidity (RH) is also high. However, if the air is near to saturation point (100 percent RH) then condensation will occur on the carcass surface, favouring mould and bacteria growth. A compromise between the two problems seems to be an RH of about 90 percent with an air speed of about 0.5 m/second. Condensation will also occur if warm carcasses are put in a cooler partially filled with cold carcasses.

The cooler should not be overloaded beyond the maximum load specified by the manufacturers and spaces should be left between carcasses for the cold air to circulate. Otherwise cooling will be inefficient and the carcass surface will remain wet, favouring rapid bacterial growth forming slime.

Once filled, a cooler should be closed and the door opened as little as possible to avoid sudden rises in temperature. When emptied, it should be thoroughly washed before refilling. Personnel handling carcasses during loading and unloading operations should follow the strictest rules regarding their personal hygiene and clothing and should handle carcasses as little as possible.

Marketing of Meat under Refrigeration

Chilled meat must be kept cold until it is sold or cooked. If the cold chain is broken, condensation forms and microbes grow rapidly. The same rules about not overloading, leaving space for air circulation, opening doors as little as possible and observing the highest hygiene standards when handling the meat apply. An ideal storage temperature for fresh meat is just above its freezing point, which is about - 1°C (- 3°C for bacon because of the presence of salt). The expected storage life given by the International Institute of Refrigeration of various types of meat held at these temperatures is as follows:

Type of meat	*Expected storage life at - 1°C*
Beef	up to 3 weeks (4–5 with strict hygiene)
Veal	1–3 weeks
Lamb	10–15 days
Pork	1–2 weeks
Edible offal	7 days
Rabbit	5 days
Bacon	4 weeks (at - 3°C)

Under commercial conditions, meat temperatures are rarely kept at - 1°C to 0°C, so actual storage times are less than expected. The times would also be reduced if RH were greater than 90 percent.

Meat should be placed in the refrigerator immediately following receipt. Any parts which show signs of mould growth or bacterial slime should be trimmed off and destroyed. Hands must be thoroughly washed after handling such trimmings and knives must be sterilized in boiling water. The refrigerator should be thoroughly cleaned after finding such meat and should also be cleaned on a regular basis.

Carcasses, quarters and large primals should not be cut into smaller portions before it is necessary as this will expose a greater surface area for bacteria to grow. Freshly cut surfaces are moist and provide a better medium for bacterial growth than the desiccated outer surfaces of cuts that have been stored for some time.

An accurate thermometer should be placed in the refrigerator and checked regularly. The temperature should remain within a narrow range (0° to + 1°C).

Transport of Meat

Vehicles for transporting meat and carcasses should be considered as an extension of the refrigerated storage. The object must be to maintain the meat temperature at or near 0°C. Meat should be chilled to 0°C before loading. Meat should hang on rails, not on the floor. If stockinettes are put on carcasses they must be clean. Meat trucks should not carry anything other than meat.

The refrigeration is usually produced by injecting liquid nitrogen or carbon dioxide (CO_2) into the compartment or by blowing air over CO_2 chunks (dry ice). The temperature in these vans can be set and controlled to minimize the temperature rise and to avoid condensation on the meat surface.

Insulated vans without refrigeration may be refrigerated by adding dry ice. While this is a reasonably good alternative to the refrigerated truck it does not allow the temperature to be controlled.

Uninsulated vans and open trucks should not be considered as suitable transport for meat, particularly in hot climates. In addition to the temperature abuse, condensation will occur when the meat goes back into refrigeration, and in open trucks the meat is exposed to attack from insects. Loading and unloading should be done quickly. If there are any unavoidable delays then dry-ice blocks should be placed in the partly filled van.

Meat Handling and Marketing without Refrigeration

Where refrigeration is unavailable either owing to financial or technical reasons (e.g. no power supply), the shelf-life of meat is reduced to days or hours, not weeks. Slaughter and dressing must be near the point of sale and it must be quick and clean. If carcasses and meat are kept in well-insulated rooms, the temperature can be reduced with dry-ice blocks, if these are available. Since it is easier to chill boneless cuts rather than whole carcasses, hot-boning should be considered.

Stock must be handled carefully to avoid producing high-pH meat which will spoil more quickly. Rooms used for slaughter and handling meat must be clean and well ventilated, but out of direct sunlight, dust-free and verminfree (rodents and insects). Hot water (82°C) must be available to clean all equipment and surfaces and personnel must work very hygienically. Receive all blood into sealed containers and have separate skips on wheels for hooves, skins, green offal and trimmings.

Dressing on a vertical hoist will minimize contamination by floor or cradle contact. Let nothing drop on the floor, only into skips. Personal hygiene must be scrupulous. Any spills of gut contents on to the meat should be cut off, but careful work will avoid this. The dressed carcass should be hung on rails. If beef is quartered to facilitate handling, the cut surface is at risk.

Red offal should be hung on hooks. Any offal processing must be in rooms away from meat-handling facilities. Intestines for human consumption must be thoroughly cleaned and washed.

Storage and Transport without Refrigeration

Meat should be put on sale within a day of slaughter. If it has to be held it should be hung in a clean, well-lit hall with good ventilation. Insects, rodents and birds must be kept out, dust must not blow in. Trays of offal should be on shelves, not on the floor. Barrows for wheeling carcasses and quarters are better than carrying on shoulders, as they can be cleaned frequently. All staff must wear clean clothing and observe strict personal hygiene. Transport of non-refrigerated meat is very hazardous. If meat is to be put in stockinettes and sacks these must be very clean. Meat should be on rails in the truck or wagon, and it is not advisable to carry it more than a day's journey before sale.

References

Grandin, T. "Best Practices for Animal Handling and Stunning", *Meat & Poultry*, April 2000, pg. 76.

Grandin, T. and Deesing, M. (2008). Humane Livestock Handling. Storey Publishing, North Adams, MA, USA.

Eisnitz, Gail A. (1997), *Slaughterhouse*. Prometheus Books, cited in Torres, Bob. *Making a Killing*. AK Press, 2007, p. 46.

"Slaughterhouses", Global Action Network, accessed March 18, 2008.

"U.S. Beef and Cattle Industry", United States Department of Agriculture, cited in Torres, Bob. *Making a Killing*. AK Press, 2007, p. 45.

Bibliography

Adeyanju, S.A., Ilori J.O. & Osuntogun, C.A. *Poultry Farming in Ondo State*, Monograph, Ile-Ife, Nigeria. Department of Animal Science, Obafemi Awolowo University. 20 pp.

Agricultural Information Centre. *Livestock development technical handbook.* Nairobi, Kenya.

Arnold, Wayne (2007). "In a growing world, milk is the new oil". *The New York Times.*

Baksh, I. (1994). Permaculture for Poultry. *African Farming,* (Jan- Feb), 13-14.

Bakshi, M.P.S. & Wadhwa, M. (2011). Nutritional status of dairy animals in different regions of Punjab State in India. *Indian J. Anim. Sci.*, 81: 60–67.

BIS. (2002|). *Mineral mixture for supplementing cattle feeds – specification.* Bureau of Indian Standards, IS 1664:2002, New Delhi.

Church, D.E. (1984). *Livestock Feeds and Feeding.* Corvallis: O&B Books.

Code-EFABAR. C*ode of Good Practice for Farm Animal Breeding and Reproduction* (FOOD-CT-2003-506506). www.code-efabar.org

Devendra, C. & Fuller, M.F. (1979). *Pig production in the Tropics.* Oxford University Press.

Eisnitz, Gail A. (1997), *Slaughterhouse.* Prometheus Books, cited in Torres, Bob. *Making a Killing.* AK Press, 2007, p. 46.

Evans, Gavin (2008). "N.Z. Forecasts Fall in Dairy Prices Through 2009". *Bloomberg.com.* Retrieved 2008-09-26.

FAO. (2003). *Biosecurity in Food and Agriculture.* Discussion Paper. Committee on Agriculture, 17th Session, Rome.

Flegel, T.W. (1988). An overview of biological treatment of crop residues. *In* Kiran Singh and J. B. Schiere, eds. *Fibrous crop residues as animal feed*, pp. 41–45, New Delhi, ICAR.

Francis, P. (1984). *Poultry production in the Tropics.* Intermediate Tropical Agriculture Series. London, Longman Group Ltd

Fraser, A.F. and D.M/ Broom. (1990). F*arm Animal Welfare and Behaviour* (3rd ed.) London: Bailliere Tindall. pp.355-356.

Garcia, L.O. & Restrepo, J.I.R. (1995). *Multinutrients block handbook.* FAO Better Farming Series No. 45, Rome, FAO

Genesis Faraday. (2004). *Selective Breeding for Aquaculture – Current Trends and Future Prospects.* Workshop report. http://www.genesis-faraday.org 7pp.

Gjedrem, T. (1997). Selective breeding to improve aquaculture production. *World Aquaculture.* 33-45

Grandin, T. "Best Practices for Animal Handling and Stunning", *Meat & Poultry*, April 2000, pg. 76.

Grandin, T. and Deesing, M. (2008). Humane Livestock Handling. Storey Publishing, North Adams, MA, USA.

Greenough, P.R. (2007). *Bovine Laminitis and Lameness: A Hands-On Approach.* Edinburgh: Saunders. p.3.

Hafez, E.S.E. (1975). *The behavior of domestic animals.* 3rd edition. Baltimore, The Williams and Wilkins Co

Hogeveen, H.,W., et al., (2001), "Milking interval, milk production and milk flow-rate in an automatic milking system", *Livestock Production Science*, Vol. 72, pp. 157–167.

Hopster, H., et al., (2002), "Stress Responses during Milking; Comparing Conventional and Automatic Milking in Primiparous Dairy Cows", *Journal of Dairy Science* Vol. 85, pp. 3206–3216.

Humphrey, J. (1980). *The classification of world livestock systems.* A study prepared for the Animal Production and Health Division. FAO. AGA/MISC/80/3.

Jahnke, H.E. (1982). *Livestock production systems and livestock development in tropical Africa.* Kiel, Germany: Kieler Wissenschaftsverlag Vauk.

Knols, B.G.J. & Takken, W., (eds.) (2007). *Emerging pests and vector-borne diseases in Europe.* Wageningen, The Netherlands, Wageningen Academic Publishers

Koster, E. (Ed.) (1996). Poultry Farming Systems for Small Commercial Farmers. *Proceedings of a Poultry Workshop held at the Animal Improvement Institute*, Irene, South Africa.

McInerney, J. (1992). Economic aspects of disease in poultry production. *Proceedings XIX World Poultry Congress*, 2: 561-566.

Midwest Plan Service. (1984). *Small farms – livestock buildings and equipment.* Ames, Iowa.

Millar, K. M., (2000), "Respect for Animal Autonomy in Bioethical Analysis: The Case of Automatic Milking Systems (AMS)",*Journal of Agricultural and Environmental Ethics*, Springer, Netherlands, Vol. 12, No. 1, pp. 41–50.

Odi, H. Diambra. (1990). State of smallholder rural poultry production in Cote d'Ivoire. *Proceedings International Seminar on Smallholder Rural Poultry Production,* 9-13 October, Thessaloniki, Greece, pp.105-116.

Robertson, A. (1976). *Handbook on animal diseases in the tropics (third edition).* British Veterinary Association. Oxfordshire, UK, Burgess and Sons, LTS.

Rossing, W. and Hogewerf, P. H., (1997), "State of the art of automatic milking systems", *Computers and electronics in agriculture*, Vol. 17, pp. 1–17.

Rushen, J., et.al.. (2008). The welfare of cattle. *Animal Welfare* Vol. 5. Berlin: Springer Verlag. pp. 21-35.

Ruthenberg, H. (1980). *Farming systems in the tropics*. Clarendon Press. Oxford.

Schukken, Y.H., Hogeveen H., and Smink, B.J., (1999), "Robotic Milking and Milk Quality, Experiences From the Netherlands",*National Mastitis Council Regional Meeting Proceedings 1999*, pp. 64–69.

"Slaughterhouses", Global Action Network, accessed March 18, 2008.

ter Horst, K. (1986). *Poultry Management in rural areas of developing countries: Matching technical demands with social facts.* Unpublished report, Department of Livestock Services, Ministry of Livestock and Fisheries, Bangladesh.

URT. (2002). *Emergency Animal Disease Surveillance and Control Programme.* Dar Es Salaam, Tanzania, Ministry of Water and Livestock Development.

"U.S. Beef and Cattle Industry", United States Department of Agriculture, cited in Torres, Bob. *Making a Killing.* AK Press, 2007, p. 45.

Van Genderen, A., de Vriend, H. (1999). The future developments in farm animal breeding and reproduction and their ethical, legal and consumer implications. Report. Project Farm Animal Breeding and Society. EU-QLRT-1999- http:// www.effab.info-publications. 130 pp.

Walli, T.K. (2009). Crop residue based densified feed block technology for improving ruminant productivity. *In* Compendium T.K. Walli ed., *Satellite Symposium on Fodder Block Technology*, pp. 67–73. New Delhi, ILDEX India.

Williamson, G. & Payne, W.J.A. (1978). *An introduction to animal husbandry in the Tropics.* 3rd edition. Tropical Agriculture Series. London, Longman Group Ltd.

Wilson, T. (1994). *Integrating livestock and crops for the sustainable use and development of tropical agricultural systems.* AGSP-FAO.

Winrock International. (1992). *Animal Agriculture in Developing Countries: Technology Dimensions.* August 1992. Morrilton, Arkansas, USA.

Index